Starting Electronics

For Jane – it's not much,
but it's mine, and it's all because of you …

Starting Electronics

Fourth Edition

Keith Brindley

AMSTERDAM • BOSTON • HEIDELBERG • LONDON
NEW YORK • OXFORD • PARIS • SAN DIEGO
SAN FRANCISCO • SINGAPORE • SYDNEY • TOKYO

Newnes is an imprint of Elsevier

Newnes is an imprint of Elsevier
The Boulevard, Langford Lane, Kidlington, Oxford, OX5 1GB
225 Wyman Street, Waltham, MA 02451, USA

First edition 1994
Second edition 1999
Third edition 2004
Fourth edition 2011

British Library Cataloguing in Publication Data
A catalogue record for this book is available from the British Library

Library of Congress Number: 2011929268

ISBN: 978-0-08-096992-3

For information on all Newnes publications
visit our website www.elsevierdirect.com

Typeset by MPS Limited, a Macmillan Company, Chennai, India
www.macmillansolutions.com

Printed and bound in the United Kingdom

11 12 13 14 10 9 8 7 6 5 4 3 2 1

For Jane – it's not much,
but it's mine, and it's all because of you …

Contents

Preface

This book originated as a collection of feature articles, previously published as a series of magazine articles in the, then, leading hobbyist electronics magazine in the UK. They were chosen for publication in book form later, not only because they were so popular with readers in their original magazine appearances, but also because they are so relevant in the field of introductory electronics – a subject area in which it is evermore difficult to find information of a technical, knowledgeable, yet understandable nature for anyone wanting to "get into" electronics. This book – hopefully – is exactly that.

Since its original publication, and with each successive new edition, I have added significant new material to make sure it is all still highly relevant and up to date. Without doubt, electronics is a rapidly moving area to study, but where this book has always been and still is firmly aimed – right at the very introduction to electronics – it manages to provide one of the best entries to the subject a reader can make.

I hope you will agree that the practical nature of the book lends itself to a self-learning experience that readers can follow in a logical and easily manageable manner. I also hope that you enjoy your journey into electronics, because even though it is a highly technical and quite demanding subject, it should also be fun.

Enjoy *Starting Electronics*! That's what I intended when I wrote it, and what I continue to believe you should do.

Keith Brindley, 2011

Chapter 1

The Very First Steps

Most people look at an electronic circuit diagram, or a printed circuit board, and have no idea what they are. One component on the board, and one little squiggle on the diagram, looks much as another. For them, electronics is a black art, practiced by weird techies, spouting untranslatable jargon and abbreviations that make absolutely no sense whatsoever to the rest of us in the real world.

But this needn't be! Electronics is not a black art – it's just a science. And like any other science – chemistry, physics, or whatever – you only need to know the rules to know what's happening. What's more, if you know the rules you're set to gain an awful lot of enjoyment from it because, unlike many sciences, electronics is a practical one, more so than just about any other science. The scientific rules that electronics is built on are few and far between, and many of them don't even have to be considered when we deal in components and circuits. Most of the things you need to know about components and the ways they can be connected together are simply mechanical and don't involve complicated formulae or theories at all.

That's why electronics is a hobby that can be immensely rewarding. Knowing just a few things, you can set about building your own circuits. You can understand how many modern electronic appliances work, and you can even design your own. I'm not saying you'll be an electronics whizz-kid, of course – it really does take a lot of studying, probably a university degree, and at least several years' experience to be that – but what I am saying is that there's lots you can do with just a little practical knowledge. That's what this book is all about – starting electronics. The rest is up to you.

WHAT YOU NEED

Obviously, you'll need some basic tools and equipment. Just exactly what these are and how much they cost depends primarily on quality. But some of these tools, as you'll see in the next few pages, are pretty reasonably priced, and well worth having. Other expensive tools and equipment that the professionals often have can usually be substituted with tools or equipment costing only a fraction of the price. So, as you'll see, electronics is not an expensive hobby. Indeed, its potential reward in terms of enjoyment and satisfaction can often be significantly greater than its cost.

In this first chapter I'll give you a rundown of all the important tools and equipment: the ones you really do need. There's also some rough guidelines to their cost, so you'll know what you'll have to pay. The tools and equipment described here, however, are the most useful ones you'll ever need and chances are you'll be using them as long as you're interested in electronics. For example, I'm still using the side-cutters I got over 20 years ago. That's got to be good value for money.

TOOLS OF THE TRADE

Talking of cutters, that's the first tool you need. There are many types of cutters but the most useful sorts are side-cutters. Generally, buy a small pair – the larger ones are OK for cutting thick wires but not for much else. In electronics most wires you want to cut are thin so, for most things, the smaller the cutters the better.

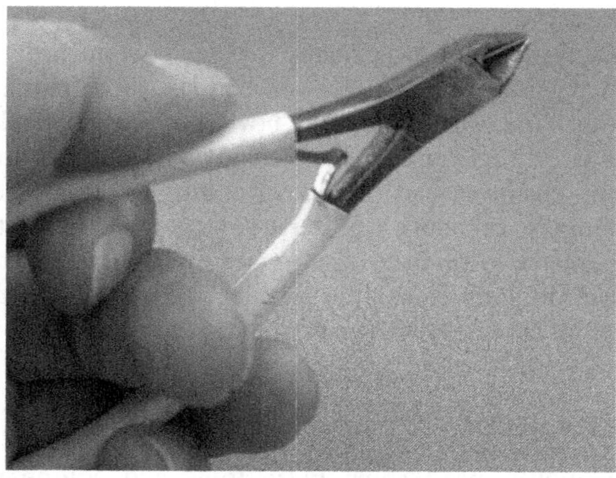

PHOTO 1.1 Side-cutters like these are essential tools – buy the best you can afford.

Hint

If you buy a small pair of side-cutters (as recommended), don't use them for cutting thick wires, or you'll find they won't last very long and you'll have wasted your money.

You can expect to pay from £4 up to about £50 or so (about US $7–85) for a good-quality pair, so look around and decide how much you want to spend.

You can use side-cutters for stripping insulation from wires too, if you're careful. But a proper wire-stripping tool makes the job much easier, and you won't cut through the wires underneath the insulation (which side-cutters are prone to do) either. There are many different types of wire strippers ranging in price from around £3 to (wait for it!) over £100 (about US $5–165). Of course, if you don't mind paying large dentist's bills you can always use your teeth – but certainly don't say I said so. You didn't hear that from me, did you?

A small pair of pliers is useful for lightly gripping components and the like. Flat-nosed or, better still, snipe-nosed varieties are preferable, costing between about £4 and £50 or so (around US $7–85). Like side-cutters, however, these are not meant for heavy-duty engineering work. Look after them and they'll look after you.

The last essential tool we're going to look at now is a soldering iron. Soldering is the process used to connect electronic

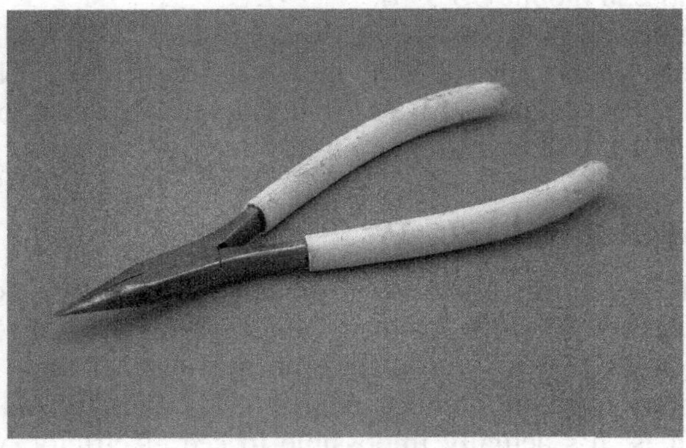

PHOTO 1.2 Snipe-nosed pliers – ideal for electronics work and another essential tool.

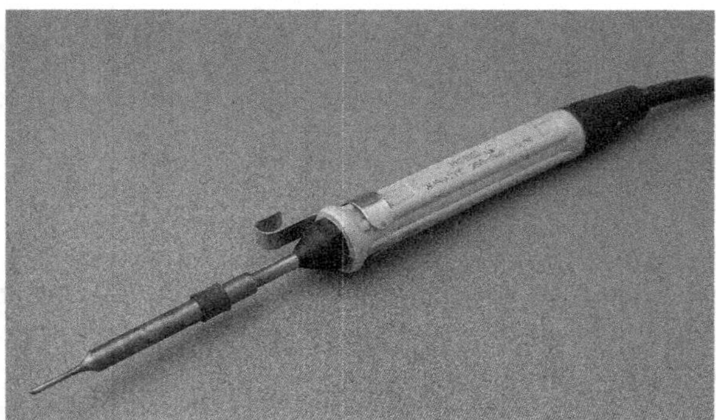

PHOTO 1.3 Low-wattage soldering iron intended for electronics.

components together, in a good permanent joint. Soldering irons range in price from about £10 to (gulp!) about £150 (about US $15–250), but – fortunately – the price doesn't necessarily reflect how useful they are in electronics. This is because irons used in electronics generally should be of pretty low power rating, because too much heat doesn't make any better a joint where tiny electronic components are concerned, and you run the risk of damaging the components too. Power rating will usually be specified on the iron or its packing and a useful iron will be around 25 watts (which may be marked 25 W).

It's possible to get soldering irons rated up to and over 100 watts, but these are of no use to you as a beginner – stick with an iron with a power rating of no more than 30–40 watts. Because of this low power need, you should be able to pick up a good iron for around £20 (about US $35).

These are all the tools we are going to look at in this chapter (I've already spent lots of your money – you'll need a breather to recover), but later on I'll be giving details of other tools and equipment that will be extremely useful to you.

IDEAS ABOUT ELECTRICITY

Electricity is a funny thing. Even though we know how to use it, how to make it do work for us, to amplify, to switch, to control, to create light or heat (you'll find out about all of these aspects of electricity over the coming chapters), we can still only guess at what it is. It's actually impossible to see electricity: we only see what it does! Sure, everyone knows that electricity is a flow

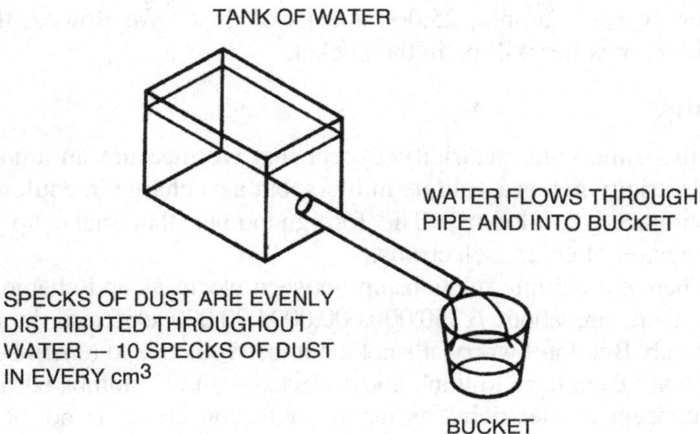
TANK OF WATER
WATER FLOWS THROUGH PIPE AND INTO BUCKET
SPECKS OF DUST ARE EVENLY DISTRIBUTED THROUGHOUT WATER — 10 SPECKS OF DUST IN EVERY cm3
BUCKET

FIGURE 1.1 Water flowing in a pipe is like electricity in a wire.

of electrons, but what are electrons? Have you ever seen one? Do you know what they look like?

The truth of the matter is that we can only hypothesize about electricity. Fortunately, the hypothesis can be seen to stand in all of the aspects of electricity and electronics we are likely to look at, so to all intents and purposes the hypothesis we have is absolute. This means we can build up ideas about electricity and be fairly sure they are correct.

Right then, let's move on to the first idea: that electricity is a flow of electrons. To put it another way, any flow of electrons is electricity. If we can measure the electricity, we must therefore be able to say how many electrons were in the flow. Think of an analogy – say, the flow of water through a pipe (Figure 1.1). The water has an evenly distributed number of foreign bodies in it. Let's say there are 10 foreign bodies (all right then, 10 specks of dust) in every cm^3 of water.

Now, if 1 liter of water pours out of the end of the pipe into the bucket shown in Figure 1.1, we can calculate the number of specks of dust that have flowed through the pipe. There's, as near as dammit, 1000 cm^3 of water in a liter, so:

$$10 \times 1000 = 10{,}000$$

water-borne specks of dust must have flowed through the pipe.

Alternatively, by knowing the number of specks of dust which have flowed through the pipe, we can calculate the volume of water. If, for example, 25,000 specks of dust have flowed, then 2.5 liters of water will be in the bucket.

Charge

It's the same with electricity, except that we measure an amount of electricity not as a volume in liters, but as a charge in coulombs (pronounced koo-looms). The foreign bodies that make up the charge are, of course, electrons.

There's a definite relationship between electrons and charge: in fact, there are about 6,250,000,000,000,000,000 electrons in one coulomb. But don't worry, it's not a number you have to remember – you don't even have to think about electrons and coulombs because the concept of electricity, as far as we're concerned, is not about electron flow, or volumes of electrons, but about flow rate and flow pressure. And as you'll now see, electricity flow rate and pressure are given their own names which – thankfully – don't even refer to electrons or coulombs. Going back to the water and pipe analogy, flow rate would be measured as a volume of water that flowed through the pipe during a defined period of time, say 10 liters in 1 minute, 1000 liters in 1 hour, or 1 liter in 1 second.

With electricity, flow rate is measured in a similar way, as a volume that flows past a point during a defined period of time, except that volume is, of course, in coulombs. So, we could say that a flow rate of electricity is 10 coulombs in 1 minute, 1000 coulombs in 1 hour, or 1 coulomb in 1 second.

We could say that, but we don't! Instead, in electricity, flow rate is called current (and given the symbol I, when drawn in a diagram).

Electric current is measured in amperes (shortened to amps, or even further shortened to the unit: A), where 1 amp is defined as a quantity of 1 coulomb passing a point in 1 second.

Instead of saying 10 coulombs in 1 minute we would therefore say:

$$\frac{10}{60} \text{ coulombs per second} = 0.167 \text{ A}$$

Similarly, instead of a flow rate of 1000 coulombs in 1 hour, we say:

$$\frac{1000}{3600} \text{ coulombs per second} = 0.3 \text{ A}$$

The other important thing we need to know about electricity is flow pressure. Returning to our analogy with water and pipe,

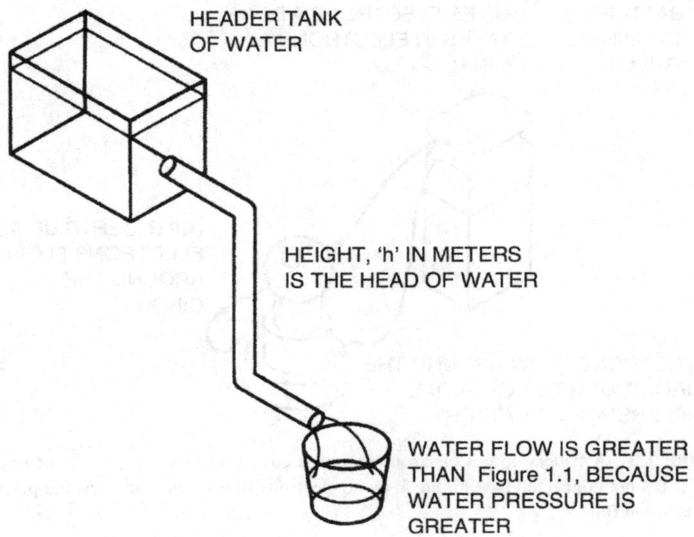
HEADER TANK
OF WATER
HEIGHT, 'h' IN METERS
IS THE HEAD OF WATER
WATER FLOW IS GREATER
THAN Figure 1.1, BECAUSE
WATER PRESSURE IS
GREATER

FIGURE 1.2 A header tank's potential energy forces the water with a higher pressure.

Figure 1.2 shows a header tank of water at a height, *h*, above the pipe. Water pressure is often classed as a head of water, where the height, *h*, in meters, is the head. The effect of gravity pushes down the water in the header tank, forming a flow pressure, forcing the water out of the pipe. It's the energy contained in the water in the header tank due to its higher position – its potential energy – that defines the water pressure.

With electricity the flow pressure is defined by the difference in numbers of electrons between two points. We say that this is a potential difference, partly because the difference depends on the positions of the points and how many electrons potentially exist. Another reason for the name potential difference comes from the early days in the pioneering of electricity, when the scientists of the day were making the first batteries. Figure 1.3 shows the basic operating principle of a battery, which simply generates electrons at one terminal and takes in electrons at the other terminal. Figure 1.3 also shows how the electrons from the battery flow around the circuit, lighting the bulb on their way round.

Under the conditions of Figure 1.4, on the other hand, nothing actually happens. This is because the two terminals aren't joined and so electrons cannot flow. (If you think about it, they are joined by air, but air is an example of a material that doesn't allow electrons

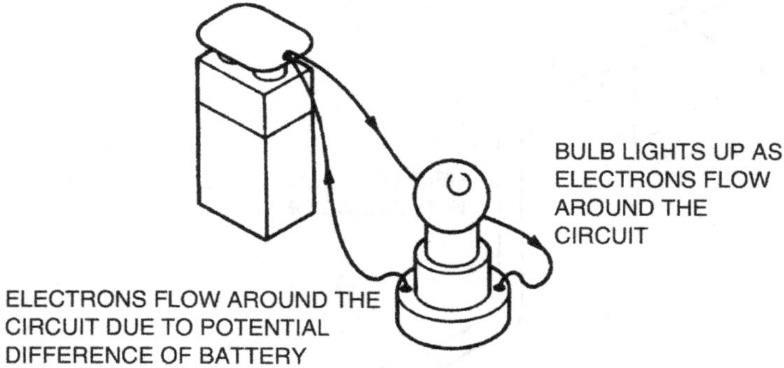
BATTERY GENERATES ELECTRONS AT ONE TERMINAL, AND TAKES IN ELECTRONS AT THE OTHER TERMINAL
BULB LIGHTS UP AS ELECTRONS FLOW AROUND THE CIRCUIT
ELECTRONS FLOW AROUND THE CIRCUIT DUE TO POTENTIAL DIFFERENCE OF BATTERY

FIGURE 1.3 A battery forces electrons around a circuit, only when the circuit is complete. If it's not connected (see Figure 1.4), no electrons flow – but it still has the potential to make them flow.

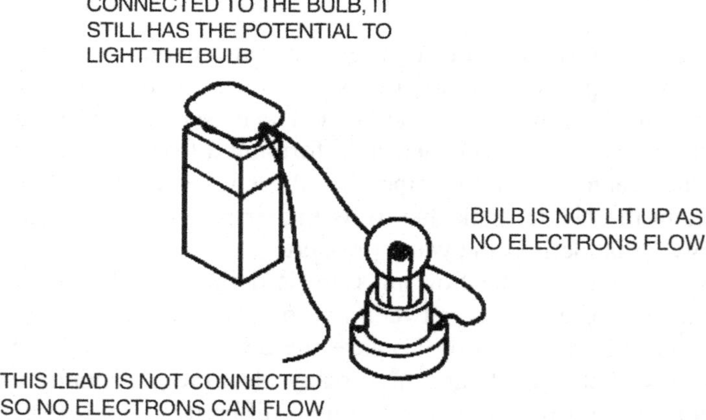
EVEN THOUGH BATTERY IS NOT CONNECTED TO THE BULB, IT STILL HAS THE POTENTIAL TO LIGHT THE BULB
BULB IS NOT LIT UP AS NO ELECTRONS FLOW
THIS LEAD IS NOT CONNECTED SO NO ELECTRONS CAN FLOW

FIGURE 1.4 Even when the battery is disconnected and electrons do not flow, the battery still has a potential difference.

to flow through it under normal conditions. Air is an insulator or a non-conductor.) Nevertheless, the battery has the potential to light the bulb and so the difference in numbers of electrons between two points (terminals in the case of a battery) is known as the potential difference. A more usual name for potential difference, though, is voltage, shortened to volts, or even the symbol V. Individual cells

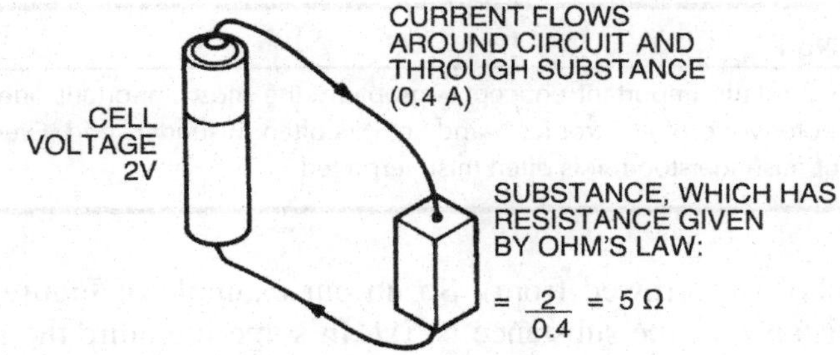
CURRENT FLOWS
AROUND CIRCUIT AND
THROUGH SUBSTANCE
(0.4 A)
CELL
VOLTAGE
2V
SUBSTANCE, WHICH HAS
RESISTANCE GIVEN
BY OHM'S LAW:
= 2/0.4 = 5 Ω

FIGURE 1.5 Cell's voltage is 2 V, and a current of 0.4 A flows.

are rated in volts and so a cell having a voltage of 3 V has a greater potential difference than a cell having a voltage of 2 V. The higher the voltage, the harder a cell can force electrons around a circuit. Voltage is simply a way of expressing electrical pushing power.

RELATIONSHIPS

You'd be right in thinking that there must be some form of relationship between this pushing power in volts and the rate of electron flow in amps. After all, the higher the voltage, the more pushing power the electrons have behind them, so the faster they should flow. The relationship was first discovered by a scientist called Ohm, and so is commonly known as Ohm's law. It may be summarized by the expression:

$$\frac{V}{I} = \text{a constant}$$

where the constant depends on the substance through which the current flows and the voltage is applied across. Figure 1.5 gives an example of a substance that is connected to a cell. The cell has a voltage of 2 V, so the voltage applied across the substance is also 2 V. The current through the substance is, in this case, 0.4 A. This means, from Ohm's law, that the constant for the substance is:

$$\frac{2}{0.4} = 5$$

The constant is commonly called the substance's resistance (because it is, in fact, a measure of the amount the substance resists the flow of current through it) and is given the unit Ω (pronounced ohm – not omega – after the scientist, not the Greek letter

> **Take Note**
>
> This is a vitally important concept – probably the most important one in the whole world of electronics – and yet it is often misunderstood. Even if it is not misunderstood, it is often misinterpreted.

its symbol is borrowed from). So, in our example of Figure 1.5, the resistance of the substance is 5 Ω. In some literature the letter R is used instead of Ω. Different substances may have different resistances and may therefore change the current flowing.

Indeed, this is so important, let's recap it and see what it all means.

If a voltage (V – measured in volts) is applied across a resistance (R – measured in ohms), a current (I – measured in amps) will flow. The voltage, current, and resistance are related by the expression:

$$\frac{V}{I} = R \tag{1.1}$$

The importance of this is that the current that flows depends entirely on the values of the resistance and the voltage. The value of the current may be determined simply by rearranging expression (1.1), to give:

$$\frac{V}{R} = I \tag{1.2}$$

So, a voltage of say 10V, applied across a resistance of 20 Ω, produces a current of:

$$\frac{10}{20} = 0.5\ \text{A}$$

Similarly, if we have a resistance, and a current is made to flow through it, then a voltage is produced across it. The value of the voltage may be determined by again rearranging expression (1.1), so that it now gives:

$$V = IR \tag{1.3}$$

Thus, a current of, say, 1 A flowing through a resistance of 5 Ω, produces a voltage of:

$$1 \times 5 = 5\ \text{V}$$

across the resistance.

Hint

Ohm's law

A simple method to help you remember Ohm's law: remember a triangle, divided into three parts. Voltage (*V*) is at the top. Current (*I*) and resistance (*R*) are at the bottom: it doesn't matter which way round *I* and *R* are – the important thing to remember is *V* at the top.

Then, if you have any two of the constants, cover up the missing one with your finger and the formula for calculating the missing one will appear. Say you know the voltage across a resistor and the current through it, but you need to know the resistance itself. Simply cover the letter *R* with your finger:

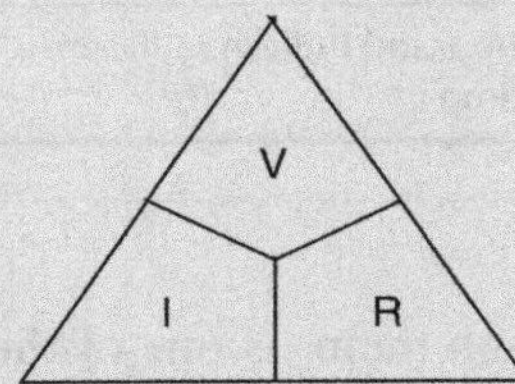

and the formula to calculate the resistance is then given as:

$$\frac{V}{I}$$

These three expressions which combine to make Ohm's law are the most common ones you'll ever meet in electronics, so look at 'em, read 'em, use 'em, learn 'em, inwardly digest 'em – just don't forget 'em. Right? Right.

Take Note

And another thing. See the way we've said throughout that a voltage is applied or produced across a resistance. Similarly a current flows through a resistance. Well let's keep it like that! Huh? Just remember that a voltage is across: a voltage does not flow through. Likewise, a current flows through: it is not across.

There is no such thing as a flow of voltage through a resistance, and there's no such thing as a current across a resistance.

ELECTRONIC COMPONENTS

The fact that different resistances produce different currents if a voltage is applied across them, or produce different voltages if a

TABLE 1.1 Comparing Amps with Smaller Values of Current

Current Name	Meaning	Value	Symbol
Amp	–	10^{0} A	A
Milliamp	One thousandth of an amp	10^{-3} A	mA
Microamp	One millionth of an amp	10^{-6} A	μA
Nanoamp	One thousand millionth of an amp	10^{-9} A	nA
Picoamp	One million millionth of an amp	10^{-12} A	pA
Femtoamp	One thousand million millionth of an amp	10^{-15} A	fA

current is applied through them, is one of the most useful facts in electronics.

In electronics, an amp of current is very large – usually we only use much smaller currents, say, a thousandth or so of an amp. Sometimes we even use currents smaller than this – say, a millionth of an amp! Similarly, we sometimes need only small voltages too.

Resistances are extremely useful in these cases, because they can be used to reduce the current flow or the voltage produced across them, due to the effects of Ohm's law. We'll look at ways and means of doing this in the next chapter. All we need to know for now is that resistances are used in electronics to control current and voltage.

Table 1.1 shows how amps are related to the smaller values of current. A thousandth of an amp is known as a milliamp.

Even smaller values of current are possible: a thousand millionth of an amp is a nanoamp (unit: nA); a million millionth is a picoamp (unit: pA). Chances are, you will never use or even specify a current value smaller than these, and you will rarely even use picoamp. Milliamps and microamps are quite commonly used, though.

It's easy to move from one current value range to another, simply by moving the decimal point one way or the other by the correct multiple of three decimal places. In this way, a current of 0.01 mA is the same as a current of 10 μA, which is the same as a current of 10,000 nA, and so on.

TABLE 1.2 Comparing Volts with Smaller and Larger Voltages

Voltage Name	Meaning	Value	Symbol
Megavolt	One million volts	10^6 V	MV
Kilovolt	One thousand volts	10^3 V	kV
Volt	–	10^0 V	V
Millivolt	One thousandth of a volt	10^{-3} V	mV
Microvolt	One millionth of a volt	10^{-6} V	μV
Nanovolt	One thousand millionth of a volt	10^{-9} V	nV

Table 1.2 shows, similarly, how volts are related to smaller values of voltage. Sometimes, however, large voltages exist (not so much in electronics, but in power electricity) and so these have been included in the table. The smaller values correspond to those of current – that is, a thousandth of a volt is a millivolt (unit: mV), a millionth of a volt is a microvolt (unit: μV), and so on – although anything smaller than a millivolt is, again, only rarely used.

Larger values of voltage are the kilovolt (that is, one thousand volts; unit: kV) and the megavolt (that is, one million volts; unit: MV). In electronics, however, these are never used.

RESISTORS

The components that are used as resistances are called, naturally enough, resistors. So that we can control current and voltage in specified ways, resistors are available in a number of values. Obviously, it would be impractical to have resistors of every possible value (for example, 1 Ω, 2 Ω, 3 Ω, 4 Ω) because literally hundreds of thousands – if not millions – of values would have to exist.

Instead, agreed ranges of values exist, and manufacturers make their resistors to have those values, within a certain tolerance. Table 1.3 shows a typical range of resistor values. This range is the most common. You can see from it that large values of resistors are available, measured in kilohms (that is, thousands of ohms; unit: kΩ) and even megohms (that is, millions of ohms; unit: MΩ). Sometimes the unit Ω (or the letter R if used) is omitted, leaving the units as just k or M.

TABLE 1.3 Typical Resistor Value Range

1Ω	10Ω	100Ω	1k	10k	100k	1M	10M
1.2Ω	12Ω	120Ω	1k2	12k	120k	1M2	–
1.5Ω	15Ω	150Ω	1k5	15k	150k	1M5	–
1.8Ω	18Ω	180Ω	1k8	18k	180k	1M8	–
2.2Ω	22Ω	220Ω	2k2	22k	220k	2M2	–
2.7Ω	27Ω	270Ω	2k7	27k	270k	2M7	–
3.3Ω	33Ω	330Ω	3k3	33k	330k	3M3	–
3.9Ω	39Ω	390Ω	3k9	39k	390k	3M9	–
4.7Ω	47Ω	470Ω	4k7	47k	470k	4M7	–
5.6Ω	56Ω	560Ω	5k6	56k	560k	5M6	–
6.8Ω	68Ω	680Ω	6k8	68k	680k	6M8	–
8.2Ω	82Ω	820Ω	8k2	82k	820k	8M2	–

Note: The decimal point in a component's value is often replaced by the multiplier e.g. 1k2 is 1200.

Resistor tolerance is specified as a plus or minus percentage. A 10Ω ± 10% resistor, say, may have an actual resistance within the range 10Ω – 10% to 10Ω + 10% – that is, between 9 and 11Ω.

As well as being rated in value and tolerance, resistors are also rated by the amount of power they can safely dissipate as heat, without being damaged. As you'll remember from our discussion on soldering irons earlier, power rating is expressed in watts (unit: W), and this is true of resistor power ratings too.

As the currents and voltages we use in electronics are normally quite small, the resistors we use also have small power ratings. Typical everyday resistors have ratings of 1/4W, 1/3W, 1/2W, 1W and so on. At the other end of the scale, for use in power electrical work, resistors are available with power ratings up to and over 100W or so.

The choice of resistor power rating you need depends on the resistor's use, but a reasonable value for electronics use is 1/4W. In fact, 1/4W is such a common power rating for a resistor that you can assume it for all the circuits in this book. If I give you a circuit to build that uses resistors of different power ratings, I'll tell you.

TIME OUT

That's all we're going to say about resistors here – in this chapter at least. In the next chapter, though, we'll be taking a look at some simple circuits you can build with resistors. We'll also explain how to measure electricity with the aid of a meter, another useful tool that is so often used in electronics. But that's enough for now, you've learned a lot in only a little time.

If, on the other hand, you feel you want to test your brain a bit more, try the quiz that follows.

QUIZ

Answers at the end of the book.

1. 100 coulombs of electricity flow past a point in an electrical circuit in 20 seconds. The current flowing is:
   a. 10 A
   b. 2 A
   c. 5 V
   d. 5 A
   e. None of these.
2. A resistor of value 1 kΩ is placed in a simple circuit with a battery of 15 V potential difference. What is the value of current that flows?:
   a. 15 mA
   b. 150 mA
   c. 1.5 mA
   d. 66.7 mA
   e. None of these.
3. A voltage of 20 V is applied across a resistor of 100 Ω. What happens?:
   a. A current of 0.2 A is generated across the resistor
   b. A current of 5 A is generated across the resistor
   c. A current of 5 A flows through the resistor
   d. One coulomb of electricity flows through the resistor
   e. None of these.
4. A current of 1 A flows through a resistor of 10 Ω. What voltage is produced through the resistor?:
   a. 10 V
   b. 1 V
   c. 100 V
   d. 10 C
   e. None of these.
5. A nanoamp is:
   a. 1×10^{-6} A
   b. 1×10^{-8} A
   c. 1000×10^{-12} A

d. 1000×10^{-6}A
e. None of these.

6. A voltage of 10 MV is applied across a resistor of 1 MΩ. What is the current that flows?:
a. 10 μA
b. 10 mA
c. 10 A
d. 10 MA
e. None of these.

Chapter 2

On the Boards

In this chapter we give you details of some easy-to-do experiments, designed to give you valuable practical experience. To perform these experiments you'll need some simple components and a couple of new tools.

The components you need are:

- 2 × 10k resistors
- 2 × 1k5 resistors
- 2 × 150 Ω resistors.

Power ratings and tolerances of these resistors are not important – just get the cheapest you can find.

The tools, on the other hand:

- a breadboard (such as the one we use)
- a multimeter (such as the multimeter we use)

are important. They are, unfortunately, quite expensive but if looked after will last you a long, long time. They're worth the expense, because you'll be able to use them for all your experiments and projects that you do and build.

In the last chapter we looked at some of the essential tools you'll need if you intend to progress very far in electronics. Breadboards and multimeters are two more that are also very much essential if you're at all serious in your intent to learn about electronics.

Fortunately, all the tools we show you in this book will last for years if properly treated, so even though it may seem like a lot of weeks' pension money now, it's money well spent as it's well worth getting the best you can afford!

ALL ABOARD

A breadboard is extremely useful. With a breadboard you can construct circuits in a temporary form, changing components if

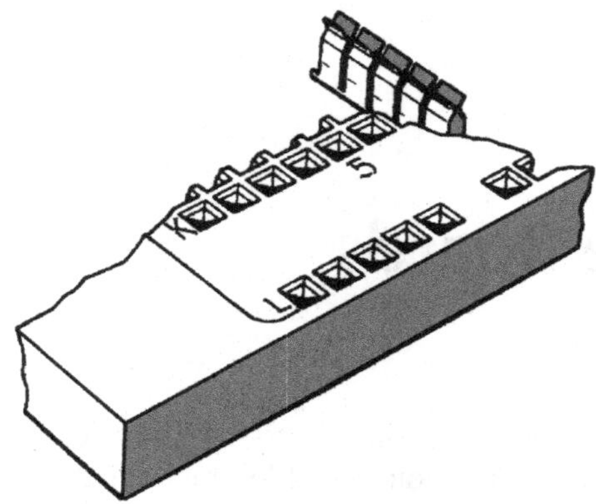

FIGURE 2.1 The interior of a breadboard, showing the contacts.

required, before committing them to a permanent circuit board. This is of most benefit if you are designing the circuit from scratch and have to change components often.

If you are following a book like *Starting Electronics*, however, a breadboard is even more useful. This is because the many circuits given in the book can be built up experimentally, tested, then dismantled, so that the components may be used again and again. I'll be giving you many such experimental circuits and, although I'll also give you good descriptions of the circuits, there's nothing like building it yourself to find out how a circuit works. So, I recommend you get the best kind of breadboard you can find – it's worth it in the long run.

There are many varieties of breadboard. All of the better ones consist basically of a moulded plastic body that has a number of holes in the top surface, through which component leads may be easily pushed. Underneath each hole is a clip mechanism, which holds the component lead tight enough so that it can't fall out. Figure 2.1 gives the idea. The clip forms a good electrical contact, yet allows the lead to be pulled out without damage.

Generally, the clips are interconnected in groups, so that by pushing leads of two different components into two holes of one group you have made an electrical contact between the two leads. In this way the component leads don't have to physically touch above the surface of the breadboard to make electrical contact.

Differences lie between breadboards in the spacings and positionings of the holes, and the number of holes in each group.

Hint

The theoretical breadboard used for artwork layout purposes in this book is specified with a total of 550 contacts arranged in a main matrix of two blocks of 47 rows of five interconnected sockets, and a row of 40 interconnected sockets down each side of the main matrix. Such a theoretical block will hold up to six 14-pin or nine eight-pin dual-in-line integrated circuit packages, together with their ancillary components.

Remember, though, that you may need to adapt the theoretical circuit layout to suit whichever particular breadboard you buy.

The majority of breadboards have hole spacings of about 2.5 mm (actually 0.1 in., which is the exact hole spacing required by a particular type of electronic component: the dual-in-line integrated circuit – I'll talk about this soon), which is fine for general-purpose use, so the only things you have to choose between are the numbers of holes in groups, the size of the breadboard, and the layout (that is, where the groups are) on the breadboard.

Because there are so many different types of breadboard available, we don't specify a standard type to use in this book. So the choice of what to buy is up to you. We do, however, show circuits on a basic breadboard, which is a fairly common layout. So, any circuits we show you to build on this breadboard can also be built on any similar quality breadboard, but you may have to adjust the actual practical circuit layout to suit.

Photo 2.1 shows a photograph of the breadboard we use throughout this book, in which you can see the top surface with all the component holes. Photo 2.2 shows the inside of the breadboard, with component lead clips interconnected into groups. The groups of clips are organized as two rows, the closest holes being 7.5 mm (not just by coincidence the distance between the rows of pins of a dual-in-line integrated circuit package) apart.

ICs

Hey, wait a minute – I've mentioned a few times already this mysterious component called a dual-in-line integrated circuit, but what is it? Well, Photo 2.3 shows one in close-up while Photo 2.4 shows it, in situ, in a professional plugblock. The pins (which provide connections to the circuits integrated inside the body – integrated circuit: geddit?) are in two rows 7.5 mm apart (actually,

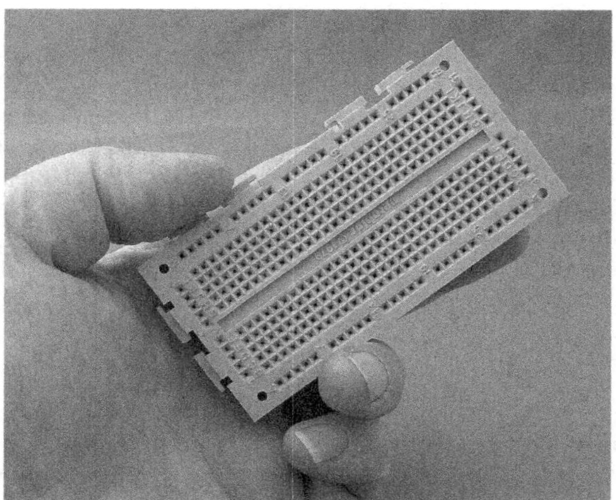

PHOTO 2.1 A typical breadboard.

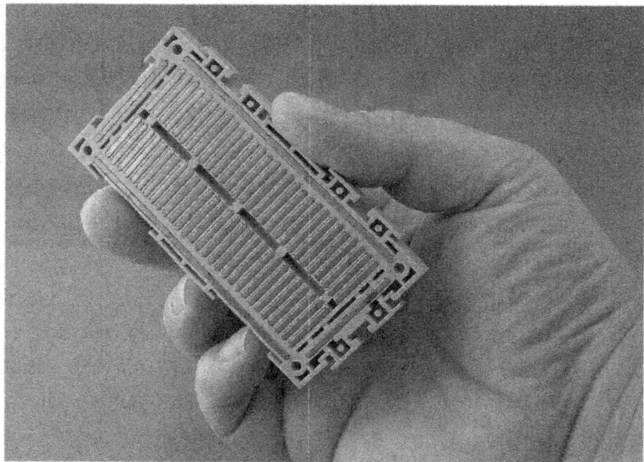

PHOTO 2.2 Inside the breadboard, showing component clips interconnected in groups.

they're exactly 0.3 in. apart), so it pushes neatly into the breadboard. Because there are two rows and they are parallel – that is, in line – we call it dual-in-line (often shortened to DIL – and while we're on the subject of abbreviations, the term dual-in-line package is often shortened to DIP, and integrated circuit also is often shortened to IC).

Many types of IC exist. Most – at least as far as the hobbyist is concerned – are in this DIL form, but other shapes do exist. Often

PHOTO 2.3 A DIL (dual-in-line) IC package.

PHOTO 2.4 An IC mounted on a breadboard. The breadboard is designed so that an IC can be mounted without shorting the pins.

DIL ICs have different numbers of pins, e.g. 8, 14, 16, 18, 28, but the pins are always in two rows. Some of the DIL ICs with large numbers of pins have rows spaced 15 mm apart (actually, exactly 0.6 in.), though.

The circuits integrated inside the body of the ICs are not always the same, and so one IC cannot automatically do the job of another. They need to be exactly the same type to be able to do that. This is why I always give a type number if I use an IC in an experiment. Make sure you buy the right one if you want to build an experiment, or for that matter if you ever build a project such as those you see in electronics magazines.

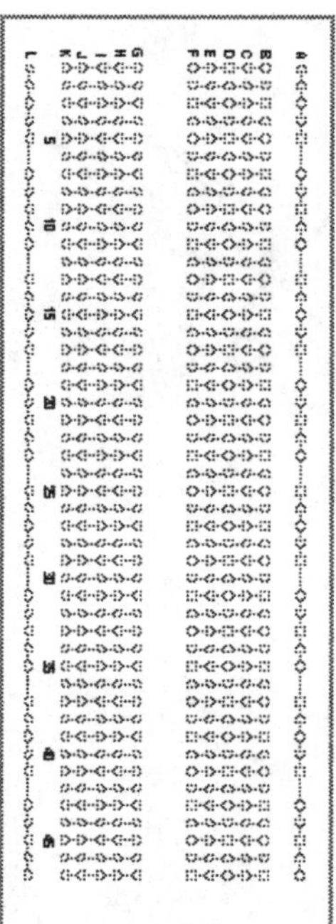

FIGURE 2.2 A breadboard pattern showing graphically the internal contacts.

Once the IC is in the breadboard – in fact, once any component is in the breadboard – it's a simple matter to make connections to it by pushing in wires or other component leads to the holes and clips of the same groups.

Down the edges of the breadboard are other groups of holes connected underneath too. These are useful to carry power supply voltages from, say, a battery, which may need to be connected into the circuit at a number of points.

We can show all the various groups of holes in the breadboard block by means of the diagram in Figure 2.2, where the connected holes are shown joined by lines. This type of diagram, incidentally, will be used throughout this book to show how the experimental circuits we look at are built using breadboard blocks. Obviously, any circuit may be built in a lot of different ways and so you don't have to follow my diagrams, or use the same breadboard as used here, but doing so will mean that your circuit is the same as mine and so easier to compare. The choice is yours. And, remember, the big advantage about using a breadboard is that the components can be pulled out when the circuit is finished and you can use them again (provided you've been careful and haven't damaged them).

THE FIRST CIRCUIT

We've done a lot of talking up to now, and not much doing, but now it's time to use your breadboard to build your first circuit. Well, to be truthful it's not really a circuit – it's just a single resistor stuck into the breadboard so that we can experiment with it.

The experiments in this chapter are all fairly simple ones, measuring the resistances of various resistors and their associated circuits. But to measure the resistances we need the other essential tool I mentioned earlier – the multimeter (Photo 2.5). Strictly speaking, a multimeter isn't just a tool used in electronics, it's a complete piece of equipment. It can be used not only to measure resistance of resistors, but also voltage and current in a circuit. Indeed, some expensive multimeters may be used to measure other things too. However, you don't need an expensive one to measure only the essentials (and some non-essentials too).

Note that any modern multimeter should meet this specification. The specification represents just the absolute minimum you should check for, and was originally drawn up for use when buying an analog multimeter (i.e. one with a pointer). Most modern multimeters are of a digital nature (i.e. with a digital readout) and so will usually greatly exceed the minimum specification.

While it's impossible for me to comment on how you intend using your multimeter, so it's impossible for me to tell you which one to buy. On the other hand, it is possible for me to recommend a few specifications that you should try to match or better when you buy your multimeter. This is simply to ensure that your multimeter will be as general-purpose as possible, and will perform measurements for you long after you progress from being a

Hint

The multimeter you buy and use is not important – as long as it meets a certain specification it will do the job nicely. This specification is discussed on page 24.

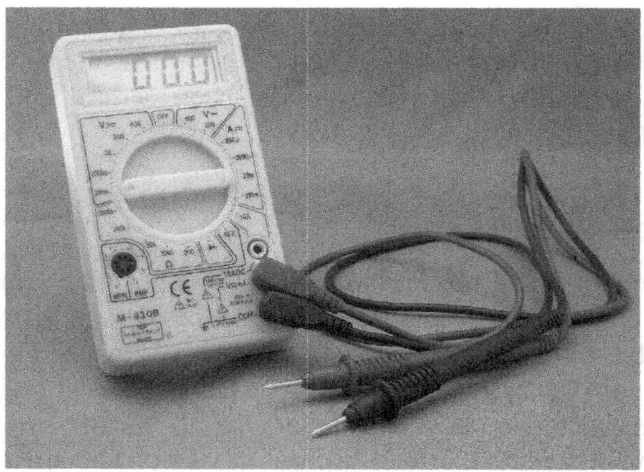

PHOTO 2.5 A multimeter – the one used throughout this book – although any multimeter with at least the specification given will do.

beginner in electronics to being an expert. The important points to remember are:

- It must have a sensitivity of at least $20\,\text{k}\Omega\,\text{V}^{-1}$ on d.c. ranges (d.c. stands for direct current)
- It must have an accuracy of no worse than $\pm$ 5%
- Its smallest d.c. voltage range should be no greater than 1 V
- Its smallest current range should be no greater than 500 μA
- It should measure resistance in at least three ranges.

In practice, just about any modern multimeter will meet and exceed this specification. Only older style analog multimeters may fall below it – digital multimeters almost always exceed it.

Using a multimeter is fairly simple. It will probably have a switch on the front, which turns so that you may select which range of measurement you want. When you have connected the multimeter up to the circuit you wish to measure (a pair of leads should be supplied with the multimeter), the readout will display the measurement or (on an older analog multimeter) the pointer of the multimeter moves and you can read off the measured value on the scale underneath the pointer. At the ends of the multimeter leads are probes that allow you to connect the multimeter to the circuit in question.

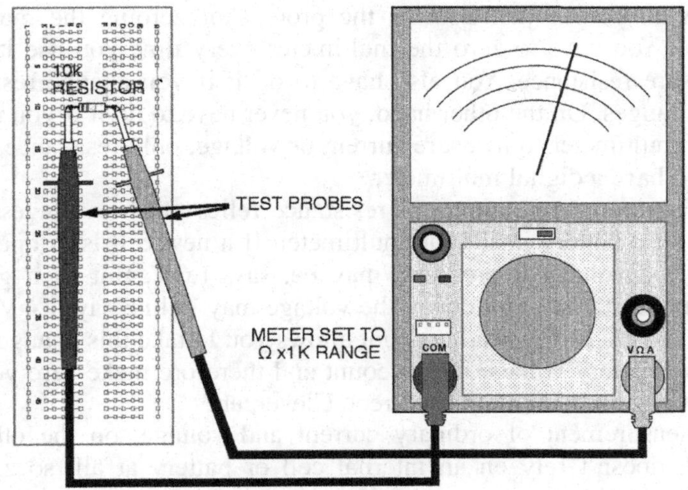

FIGURE 2.3 About the simplest circuit you could have: a single resistor and a multimeter. The multimeter takes the place of a power supply, and the circuit's job is to test the resistor!

EXPERIMENT

Using the multimeter in our first experiment – to measure a resistor's resistance – we will now go through the procedure step by step, so that you get the hang of it.

The circuit built on a breadboard is shown in Figure 2.3. Being only one resistor it's an extremely simple circuit, so simple that we are sure you would be able to do it yourself without our aid, but we might as well start off on a good footing and do the job properly – some of the circuits we'll be looking at in the following chapters will not be so simple.

If you have an analog multimeter, you have to adjust it so that the reading is accurate. The step-by-step process is as follows:

1. Turn the switch to point to a resistance range (usually marked OHM × 1 K, or similar).
2. Touch the multimeter probes together – the pointer should swing around to the right.
3. Read the resistance scale of the multimeter – the top one on our multimeter, marked OHMS, where the pointer crosses it. It should cross exactly on the number 0.
4. If it doesn't cross at 0, adjust the multimeter using the zero adjust knob (usually marked 0 ΩADJ).

What you've just done is the process of zeroing the multimeter. You have to zero the multimeter every time you use it to measure resistance. You also have to do it if you change resistance ranges. On the other hand, you never have to do it if you use your multimeter to measure current or voltage, only resistance, or if you have a digital multimeter.

You see, measurement of resistance relies on the voltages of cells or a battery inside the multimeter. If a new cell is in operation, the voltage it produces may be, say, 1.6 V. But as it gets older and starts to run down, the voltage may fall to, say, 1.4 V or even lower. The zero adjustment allows you to take this change in cell or battery voltage into account and therefore make sure your resistance measurement is correct. Clever, eh?

Measurement of ordinary current and voltage, on the other hand, doesn't rely on an internal cell or battery at all, so zero adjustment is not necessary.

Hint

Resistor Color Code

Resistance values are indicated on the bodies of resistors in one of two ways: in actual figures, or more usually by a color code. Resistors using figures are usually high-precision or high-wattage types that have sufficient space on their bodies to print characters on. Color coding, on the other hand, is the method used on the vast majority of resistors – for two reasons. First, it is easier to read when components are in place on a printed circuit board. Second, some resistors are so small it would be impossible to print numbers on them, let alone read them afterwards.

Depending on the type of resistor, the color code can be made up of four or five bands printed around the resistor's body (as shown below). The five-band code is typically used on more accurate resistors as it provides a more precise representation of value. Usually, the four-band code is adequate for most general purposes and it's the one you'll nearly always use – but you still need to be aware of both! Table 2.1 shows both resistors and lists the colors and values associated with each band of both four-band and five-band color codes.

The bands grouped together indicate the resistor's resistance value, while the single band indicates its tolerance.

The first band of the group indicates the resistor's first figure of its value. The second band is the second figure. Then, for a four-band coded resistor, the third band is the multiplier. For a five-band coded resistor the third band is simply the resistor's third figure, while the fourth band is the multiplier. For both, the multiplier is simply the factor by which the first figure should be multiplied (or simply the number of noughts to add) to obtain the actual resistance.

TABLE 2.1 Resistor Color Code

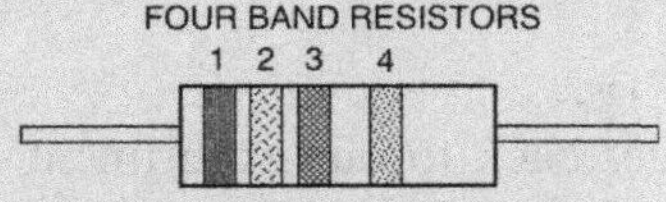

Color	Band 1 First Figure	Band 2 Second Figure	Band 3 Multiplier	Band 4 Tolerance
Black	0	0	×1	–
Brown	1	1	×10	1%
Red	2	2	×100	2%
Orange	3	3	×1000	–
Yellow	4	4	×10,000	–
Green	5	5	×100,000	–
Blue	6	6	×1,000,000	–
Violet	7	7	–	–
Gray	8	8	–	–
White	9	9	–	–
Gold	–	–	×0.1	5%
Silver	–	–	×0.1	10%

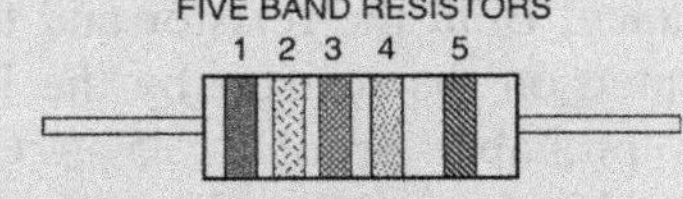

Color	Band 1 First Figure	Band 2 Second Figure	Band 3 Third Figure	Band 4 Multiplier	Band 5 Tolerance
Black	0	0	0	×1	–
Brown	1	1	1	×10	1%
Red	2	2	2	×100	2%
Orange	3	3	3	×1000	–
Yellow	4	4	4	×10,000	–
Green	5	5	5	×100,000	0.5%
Blue	6	6	6	×1,000,000	0.25%
Violet	7	7	7	×10,000,000	0.1%
Gray	8	8	8	–	0.01%
White	9	9	9	–	–
Gold	–	–	–	×0.1	5%
Silver	–	–	–	×0.1	10%

As an example, take a resistor coded red, violet, orange, silver. Looking at Table 2.1, we can see that it's obviously a four-band color-coded resistor, and its first figure is 2, the second is 7, the multiplier is ×1000, and tolerance is ±10%. In other words, its value is 27,000 Ω, or 27 k.

Now let's get back to our experiment. Following the diagram of Figure 2.3:

1. Put a 10k resistor (brown, black, orange bands) into the breadboard.
2. Touch the multimeter leads against the leads of the resistor (it doesn't matter which way round the multimeter leads are).
3. Read off the scale at the point where the pointer crosses it. What does it read? It should be 10.

But how can that be? It's a 10k resistor, isn't it? Well, the answer's simple. If you remember, you turned the multimeter's range switch to OHM ×1 K, didn't you? Officially, this should be OHM ×1 k – that is, a lower case k. This tells you that whatever reading you get on the resistance scale you multiply by 1 k – that is, 1000. So the multimeter reading is actually 10,000. And what is the value of the resistor in the breadboard – 10k (or 10,000 Ω), right!

In practice, you might find that your resistor's measurement isn't exactly 10k. It may be, say, 9.5k or 10.5k. This is due, of course, to tolerance. Both the resistor and the multimeter have a tolerance, indicated on the resistor by the last colored band: the multimeter's is probably around ±5%. Chances are, though, you'll find the multimeter reading is as close to 10k as makes no difference.

Now you've seen how your multimeter works, you can use it to measure any other resistors you have, if you wish. You'll find that lower-value resistors need to be measured with the range switch on lower ranges, say Ω × 100. Remember – if you have an analog multimeter – every time you intend to make a measurement you must first zero the multimeter. The process may seem a bit long-winded for the first two or three measurements, but after that you'll get the hang of it.

THE SECOND CIRCUIT

Figure 2.4 shows the next circuit we're going to look at and how to build it on a breadboard. It's really just another simple circuit, this time consisting of two resistors in a line – we say they're in series. The aim of this experiment is to measure the overall resistance of the series resistors and see if we can devise a formula that allows us to calculate other series resistors' overall resistances without the need of measurement.

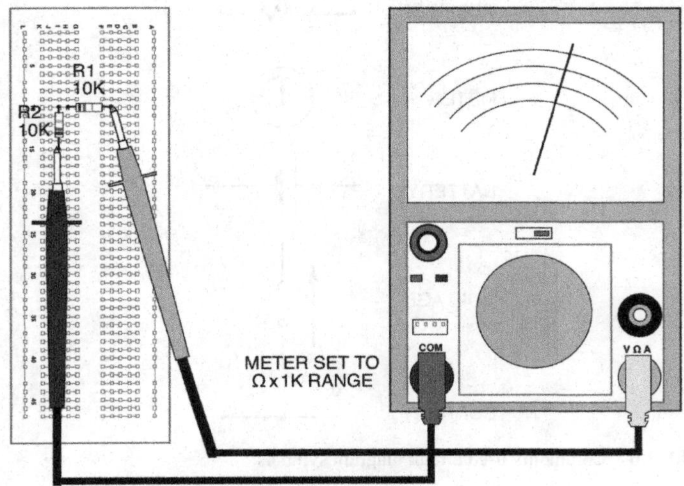

FIGURE 2.4 Two resistors mounted on the breadboard in series.

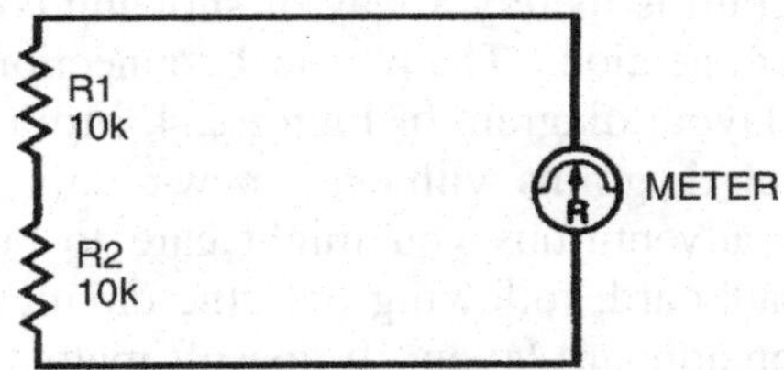

FIGURE 2.5 A circuit diagram of two resistors in series. The meter is represented by a round symbol.

Figure 2.5 shows the more usual way of representing a circuit in a drawing – the circuit diagram. What we have done is replace the actual resistor shapes with symbols. Resistor symbols are zig-zag lines usually, although sometimes small oblong boxes are used in circuit diagrams. The resistors in the circuit diagram are numbered R1 and R2, and their values are shown too.

Meters in circuit diagrams are shown as a circular symbol, with an arrow to indicate the pointer. To show it's a resistance multimeter (that is, an ohm-multimeter, more commonly called just ohmmeter) the letter R is shown inside it. While we're on the topic of circuit diagram symbols, Figure 2.6 shows a few very common ones (including resistor and meter) that we'll use in this book. Look out for them later!

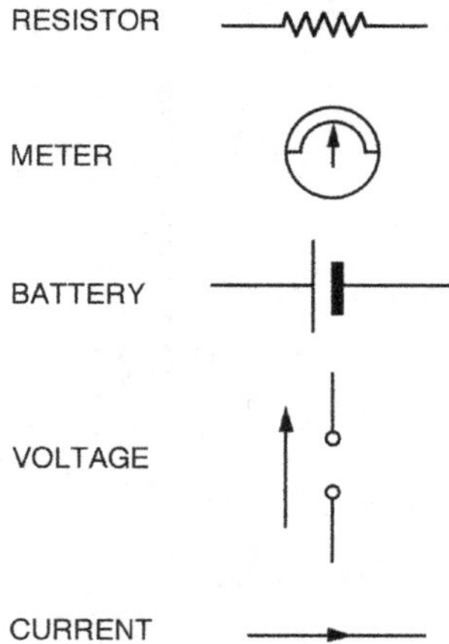

FIGURE 2.6 Commonly used circuit diagram symbols.

You should have noticed that there is no indication of the breadboard in the circuit diagram of Figure 2.5. There is no need. The circuit diagram is merely a way of showing components and their electrical connections. The physical connection details are in the breadboard layout diagram of Figure 2.4. From now on, we'll be using two such diagrams with every new circuit. If you're feeling particularly adventurous you might care to build your own circuit on a breadboard, following only the circuit diagram – not the associated breadboard layout. It doesn't matter if your circuit has a different layout to ours, it will still work as long as all the electrical connections are there.

Back to the circuit: it's now time to measure the overall resistance of the series resistors. Following the same instructions we gave you before, do it!

If your measurement is correct you should have a reading of 20k. But what does this prove? Well, it suggests that there is a relationship between the separate resistors (each of value 10k) and the overall resistance. It looks very much as though the overall resistance (which we call, say, R_{OV}) equals R1 + R2. Or put mathematically:

$$R_{OV} = \mathrm{R1} + \mathrm{R2}$$

But how can we test this? The easiest way is to change the resistors. Try doing the experiment with two different resistors. You'll find the same is true: the overall resistance always equals the sum of the two separate resistances.

By experiment, we've just proved the law of series resistors. And it doesn't just stop at two resistors in series. Three, four, five, in fact, any number of resistors may be in series – the overall resistance is the sum of the individual ones. This can be summarized mathematically as:

$$R_{OV} = R1 + R2 + R3 + \ldots$$

Try it yourself!

THE NEXT CIRCUIT

There is another way two or more resistors may be joined. Not end to end as series resistors are, but joined at both ends. We say resistors joined together at both ends are in parallel. Figure 2.7 shows the circuit diagram of two resistors joined in parallel, and Figure 2.8 shows a breadboard layout. Both these resistors are, again, 10k resistors. What do you think the overall resistance will be? It's certainly not 20k!

Measure it yourself using your multimeter and breadboard. You should find that the overall resistance is 5k. Odd, eh? Replace the two 10k resistors with resistors of a different value – say, two 150Ω resistors (brown, green, brown). The overall resistance is 75Ω.

So, we can see that if two equal-value resistors are in parallel, the overall resistance is half the value of one of them. This is a quite useful fact to remember when two parallel resistors are equal in value, but what happens when they're not?

Try the same circuit, but with unequal resistors this time – say, one of 10k and the other of 1k5 (brown, green, red – shouldn't you be learning the resistor color code?). What is the overall resistance? You should find it's about 1k3 – neither one thing nor the other! So, what's the relationship?

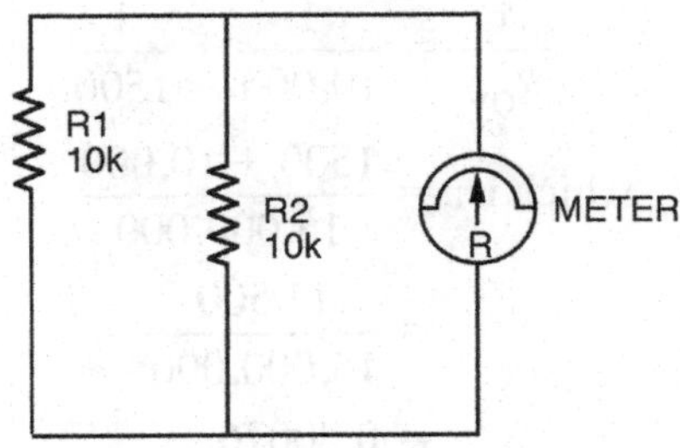

FIGURE 2.7 The circuit diagram for two resistors in parallel, with the meter symbol.

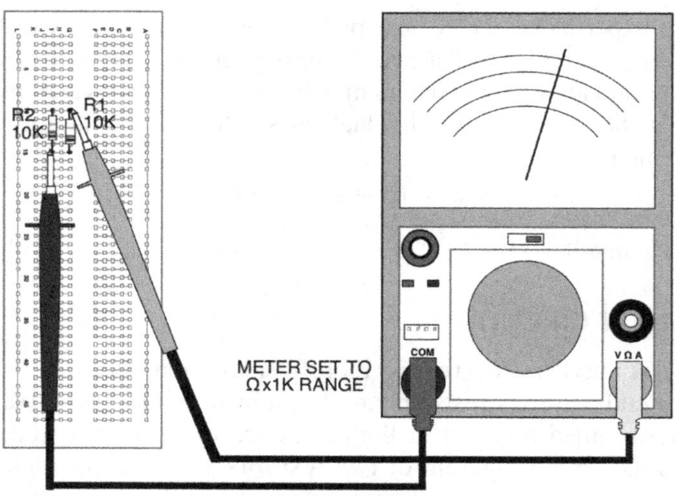

FIGURE 2.8 The two parallel resistors shown in the breadboard, with the meter in place to test their combined resistance.

Well, a clue to the relationship between parallel resistors comes from the fact that, in a funny sort of way, parallel is the inverse of series. So if we inverted the formula for series resistors we saw earlier:

$$R_{OV} = \text{R1} + \text{R2} + \text{R3} + \ldots$$

we would get:

$$\frac{1}{R_{OV}} = \frac{1}{\text{R1}} + \frac{1}{\text{R2}} + \frac{1}{\text{R3}} + \ldots$$

and this is the formula for parallel resistors. Let's try it out on the resistors of this last experiment. Putting in the values 10k and 1k5, we get:

$$\frac{1}{R_{OV}} = \frac{1}{10{,}000} + \frac{1}{1500}$$

$$\text{which is: } = \frac{1500 + 10{,}000}{15{,}000{,}000}$$

$$= \frac{11{,}500}{15{,}000{,}000}$$

$$= 0.00076$$

$$\text{So: } R_{OV} = 1304\,\Omega$$

which is about 1k3, the measured value.

This is the law of parallel resistors, every bit as important as that of series resistors. Remember it!

If there are only two resistors in parallel, you don't have to calculate it the way we've just done here. There is a simpler way, given by the expression:

$$R_{OV} = \frac{R1 \times R2}{R1 + R2}$$

But if there are three or more resistors in parallel you have to use the long method, I'm afraid.

Hint

The laws we've seen in this and the previous chapter of *Starting Electronics* (Ohm's law and the laws of series and parallel resistors) are the basic laws we need to understand all of the future things we'll look at.

MORE AND MORE COMPLEX CIRCUITS

The circuits we've looked at so far have been very simple, really. Much more complex ones await us over the coming chapters. We'll also be introduced to several new components, such as capacitors, diodes, transistors, and ICs. Fortunately, most of the new circuits and components follow these basic laws we've studied, so you'll be able to understand their operation without much ado.

The aim of these basic laws is to simplify complex circuits so that we may understand them and how they work, with reference to the circuits we have already seen. For example, the circuit in Figure 2.9 is quite complex. It consists of many resistors in a network. But by grouping the resistors into smaller circuits of only series and parallel resistors, it's possible to simplify the whole circuit into one overall resistance.

In fact, I'm going to leave you with that problem now. What you have to do is to calculate the current, *I*, which flows from the battery. You can only do that by first finding the circuit's overall resistance. If you can't do it by calculation you can always do it by building the circuit up on a breadboard and measuring it. Resistor and multimeter tolerances, of course, will mean the measured result may not give exactly the same result as the calculated one.

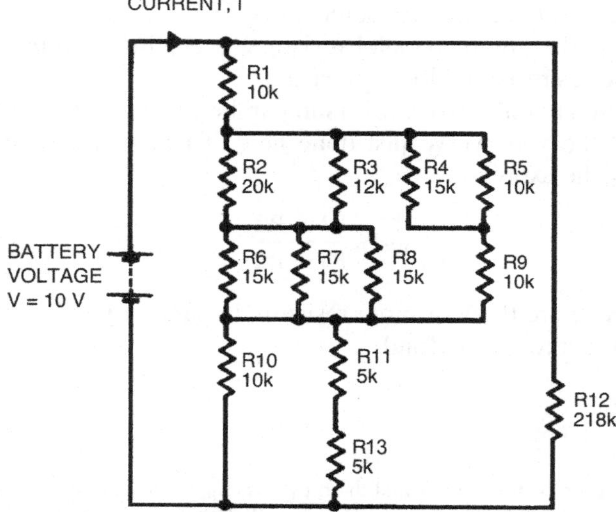

FIGURE 2.9 A comparatively complex circuit showing a number of resistors in parallel and in series. By breaking the circuit down into groups, the overall resistance can be calculated.

Meanwhile, there's always the quiz that follows to keep you occupied!

QUIZ

Answers at the end of the book.

1. Five 10kΩ resistors are in series. One 50kΩ resistor is placed in parallel across them all. The overall resistance is:
   a. 100kΩ
   b. 52kΩ
   c. 25kΩ
   d. 50kΩ
   e. None of these.
2. Three 30kΩ resistors are in parallel. The overall resistance is:
   a. 10kΩ
   b. 90kΩ
   c. 27kΩ
   d. 33kΩ
   e. 7k5
   f. None of these.

3. When must you zero a multimeter?:
   a. Whenever a new measurement is to be taken
   b. Whenever you turn the range switch
   c. Whenever you alter the zero adjust knob
   d. Whenever a resistance measurement is to be made
   e. All of these
   f. None of these.
4. The law of series resistors says that the overall resistance of a number of resistors in series is the sum of the individual resistances: true or false?
5. The overall resistance of two parallel resistors is 1 kΩ. The individual resistance of these resistors could be:
   a. 2 kΩ and 2 kΩ
   b. 3 kΩ and 1 k5
   c. 6 kΩ and 1 k2
   d. 9 kΩ and 1125 Ω
   e. a and b
   f. All of these
   g. None of these.

Chapter 3

Measuring Current and Voltage

There's not too much to buy for this chapter of *Starting Electronics*. As usual there's a small number of resistors:

- 2 × 1k5
- 1 × 4k7
- 2 × 100k.

As before, power ratings and tolerances of these resistors are not important.

We're looking at current and voltage in our experiments in this chapter, so you'll need to have a voltage and current source. The easiest and cheapest method is a simple battery. PP3, PP6, or PP9 sizes are best – as we need a 9 V source.

Whichever battery you buy, you'll also need its corresponding battery connectors. These normally come as a pair of push-on connectors and colored connecting leads. The red lead connects to the positive battery terminal; the black lead connects to the negative terminal.

The ends of the connecting leads furthest from the battery are normally stripped of insulation for about the last few millimeters or so, and tinned. Tinning is the process whereby the loose ends of the lead are soldered together. If the leads are tinned you'll be able to push them straight into your breadboard. If the leads aren't tinned, on the other hand, don't just push them in because the individual strands of wire may break off and jam up the breadboard. Instead, tin them yourself using your soldering iron and some multicore solder. The following steps should be adhered to:

1. Twist the strands of the leads between your thumb and first finger, so that they are tightly wound, with no loose strands.
2. Switch on the soldering iron. When it has heated up, tin the iron's tip with multicore solder until the end is bright and shiny with molten solder. If necessary, wipe off excess solder

or dirt from the tip, on a piece of damp sponge; keep a small piece of sponge just for this purpose.

3. Apply the tip of the iron to the lead end, and when the wires are hot enough apply the multicore solder. The solder should flow over the wires smoothly. Quickly remove the tip and the multicore solder, allowing the lead to cool naturally (don't blow on it as the solder may crack if it cools too quickly). The lead end should be covered in solder, but no excess solder should be present. If a small blob of solder has formed on the very end of the lead, preventing the lead from being pushed into your breadboard, just cut it off using your side cutters.

We'll start our study, this chapter, with a brief recap of the ideas we covered in the last chapter. We saw then that resistors in series may be considered as a single equivalent resistor, whose resistance is found by adding together the resistance of each resistor. This is called the law of series resistors, which is given mathematically by:

$$R_{OV} = \mathrm{R1} + \mathrm{R2} + \mathrm{R3} + \ldots$$

Similarly, there is a law of parallel resistors, by which the single equivalent resistance of a number of resistors connected in parallel is given by:

$$\frac{1}{R_{OV}} = \frac{1}{\mathrm{R1}} + \frac{1}{\mathrm{R2}} + \frac{1}{\mathrm{R3}} + \ldots$$

Using these two laws, many involved circuits may be broken down, step by step, into an equivalent circuit consisting of only one equivalent resistor. The circuit in Figure 2.9 in the last chapter was one such example and, if you remember, your homework was to calculate the current I from the battery. To do this you first have to find the single equivalent resistance of the whole network, then use Ohm's law to calculate the current.

Figure 3.1 shows the first stage in tackling the problem, by dividing the network up into a number of smaller networks. Using the two expressions above, associated with resistors in series and parallel, we can start to calculate the equivalent resistances of each small network as follows:

Network A

Network A consists of two resistors in parallel. The equivalent resistance, R_A, may therefore be calculated from the above

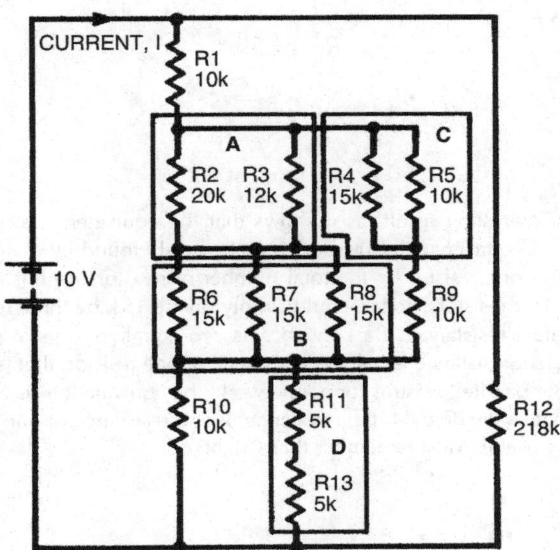

FIGURE 3.1 A resistor network in which the total resistance can only be calculated by breaking it down into blocks.

expression for parallel resistors. However, we saw in the last chapter that the resistance of only two parallel resistors is given by the much simpler expression:

$$R_A = \frac{\text{R2} \times \text{R3}}{\text{R2} + \text{R3}}$$

which gives:

$$R_A = \frac{20\text{k} \times 12\text{k}}{32\text{k}}$$
$$= 7\text{k}5$$

Network B

Three equal, parallel resistors form this network. Using the expression for parallel resistors we can calculate the equivalent resistance to be:

$$\frac{1}{R_B} = \frac{1}{15\text{k}} + \frac{1}{15\text{k}} + \frac{1}{15\text{k}} + \ldots$$

which gives:

$$R_B = 5k$$

Hint

This is an interesting result, as it shows that the equivalent resistance of a number of equal, parallel resistors may be easily found by dividing the resistance of one resistor by the total number of resistors. In this case we had three 15 k resistors, so we could simply divide 15 k by three to obtain the equivalent resistance. If a network has two equal resistors in parallel, the equivalent resistance is half the resistance of one resistor (that is, divide by 2). If four parallel resistors form a network, the equivalent resistance is a quarter (that is, divide by 4) the resistance of one resistance of one resistor. Hmm, very useful. Must remember that – right?

Network C

Using the simple expression for two, unequal parallel resistors:

$$R_C = \frac{R4 \times R5}{R4 + R5}$$
$$= 6k$$

Network D

The overall resistance of two series resistors is found by adding their individual resistances. Resistance R_D is therefore 10 k.

We can now redraw the whole network using their equivalent resistances, as in Figure 3.2, and further simplify the resultant networks.

Network E

This equivalent resistance is found by adding resistances R_A and R_B. It is therefore 12k5.

Network F

Same as network E. Resistance $R_F = 16$ k.

Network G

Two equal parallel resistors, each of 10 k. Resistance R_G is therefore 5 k. Figure 3.3 shows further simplification.

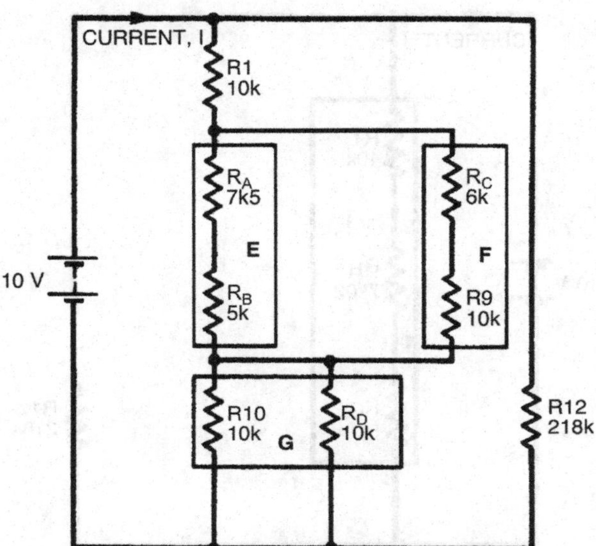

FIGURE 3.2 The same network, broken down into smaller networks using equivalent resistances (networks E, F, and G).

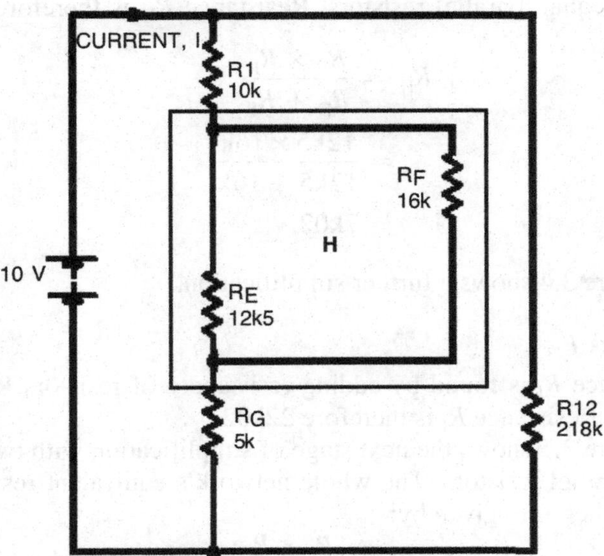

FIGURE 3.3 The network further simplified (network H) into two unequal parallel resistances.

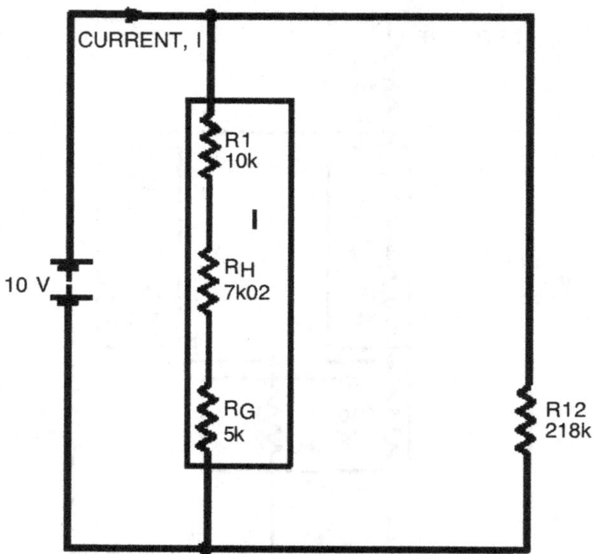

FIGURE 3.4 The network simplified into a string of series resistances (network I).

Network H

Two unequal, parallel resistors. Resistance R_H is therefore given by:

$$R_H = \frac{R_E \times R_F}{R_E + R_F} = \frac{12k5 \times 16k}{12k5 + 16k} = 7k02$$

Figure 3.4 shows a further simplification.

Network I

Resistance R_I is found by adding resistances of resistors R1, R_H, and R_G. Resistance R_I is therefore 22k02.

Figure 3.5 shows the next stage of simplification with two unequal parallel resistors. The whole network's equivalent resistance (R_N) is therefore given by:

$$R_N = \frac{R_I \times R_{12}}{R_I + R_{12}} = 20k$$

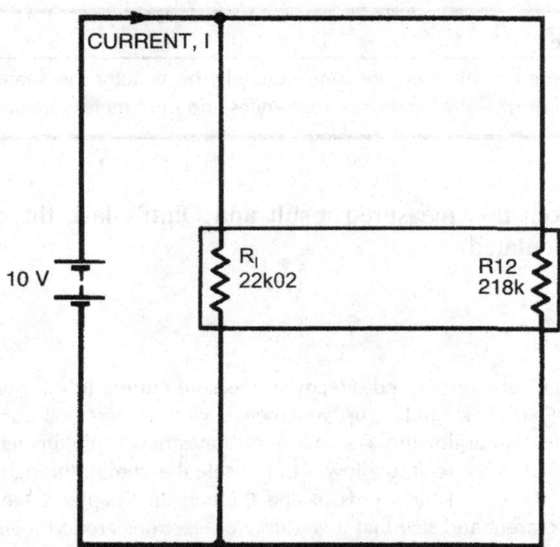

FIGURE 3.5 The final simplification, into two parallel resistances. From this the current can be calculated.

Now, from Ohm's law we may calculate the current *I*, flowing from the battery, using the expression:

$$
\begin{aligned}
I &= \frac{V}{R} \\
&= \frac{10}{20\text{k}} \\
&= 0.5 \times 10^{-3}\ \text{A} = 0.5\ \text{mA}
\end{aligned}
$$

Did you get the same answer?

ONE MAN'S METER …

Of course, there is another way the current *I* may be found. Instead of calculating the equivalent resistance of the network you could measure it using your multimeter. But first you would need to obtain resistors of all the values in the network. Then you must build the network up on a breadboard, with those resistors.

Take Note

The measured result may not, unfortunately, be exactly the same as the calculated result – due to resistor tolerances and multimeter tolerance.

Finally, from this measured result and Ohm's law, the current I may be calculated.

Hint

Your multimeter can be used directly to measure current (it's a multimeter, remember), so after building up your network of resistors you can simply read off the current flowing. Figure 3.6 illustrates how a multimeter should be connected in a circuit to allow it to indicate the current through the circuit. Note the use of the words in and through. In Chapter 1 we looked closely at current and saw that it is a flow of electrons around a circuit. To measure the current we must position the multimeter in the circuit – in other words, the current going around the circuit must also go through the multimeter. This is an important point when measuring current. Remember it!

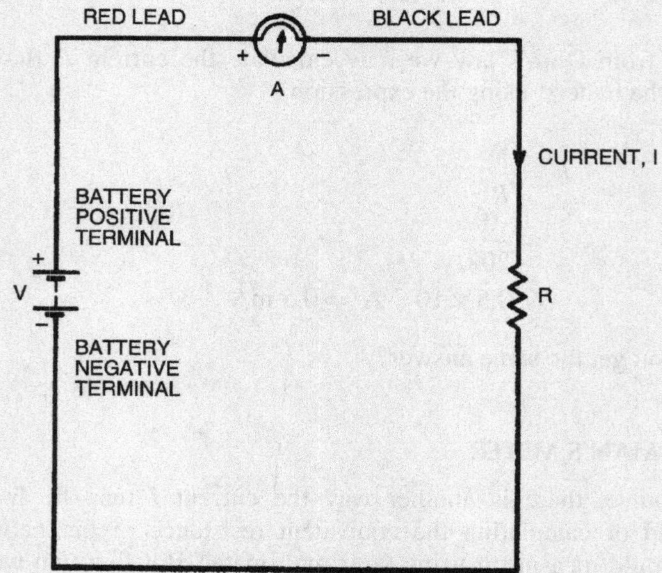

FIGURE 3.6 Diagram showing a meter connected in circuit, with the red lead at the point of higher potential.

Take Note

If you are using an analog multimeter, make sure you have your leads the correct way round. When the current through the multimeter flows in the right direction the movement and pointer turn clockwise. In the wrong direction, on the other hand, the movement and pointer attempt to turn anticlockwise. At best you won't get a valid measurement of current if the multimeter is the wrong way round – at worst you'll damage the movement.

A digital multimeter, however, will merely give a negative reading.

But what is the right way round? And how do you tell the difference? The answers are quite simple really, both given by the fact that your multimeter has two leads: one red, one black. The color coding is used to signify which lead to connect to which point in a circuit. By convention, black is taken to be the color signifying a lower potential. Red, also by convention, signifies a higher potential. So, the multimeter should be connected into the circuit with its red lead touching the point of higher potential and the black lead touching the point of lower potential. This is illustrated in Figure 3.6 with symbols close to the multimeter (+ for the red lead, − for the black). In the circuit shown, the point of higher potential is the side of the circuit close to the positive terminal of the battery (also marked +).

It doesn't matter where in the circuit the multimeter is placed, the red lead must always connect to the point of higher potential. Figure 3.7, for example, shows the multimeter at a different position, but the measurement is the same and the multimeter won't be damaged as long as the red lead is connected to the point of higher potential.

Negative Vibes

Figures 3.1–3.7 all show current flowing from the positive terminal of the battery to the negative terminal. Now we know that current is made up of a flow of electrons so we might assume that electrons also flow from positive to negative battery terminals. We might – but we'd be wrong, because electrons are actually negatively charged and therefore flow from negative to positive battery terminals. But a negative flow in one direction is exactly the same as a positive flow in the opposite direction (think about it!), so the two things mean the same. However, it does mean that we

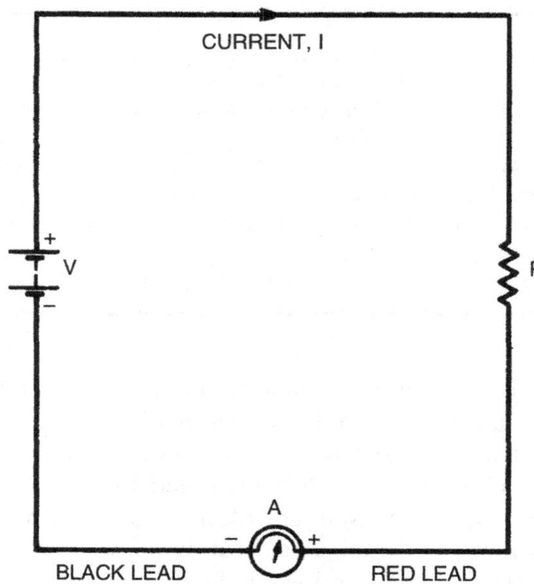

FIGURE 3.7 The meter connected to a different part of the circuit. The effect is the same as long as the red lead is correctly connected.

must define which sort of current we are talking about when we use the term. Figure 3.8 defines it graphically: conventional current (usually called simply current) flows from positive to negative; electron current flows from negative to positive. Whenever we talk about current from now on, you should take it to mean conventional current.

A METER WITH POTENTIAL

Of course, your multimeter can measure voltage too. But how should it be connected into the circuit to do it? Well, the answer is that you don't connect it into the circuit – because voltage, remember, is across a component, not through it. So, to measure the voltage across a component you must connect the multimeter across the component too. Figure 3.9 shows a simple example of a circuit comprising two resistors and a battery. To find the voltage across, say, the lower resistor, the multimeter is just connected across that resistor. Like the measurement of current, the red (+) lead of the multimeter is connected to the more positive side of the circuit and the black (−) lead to the more negative side.

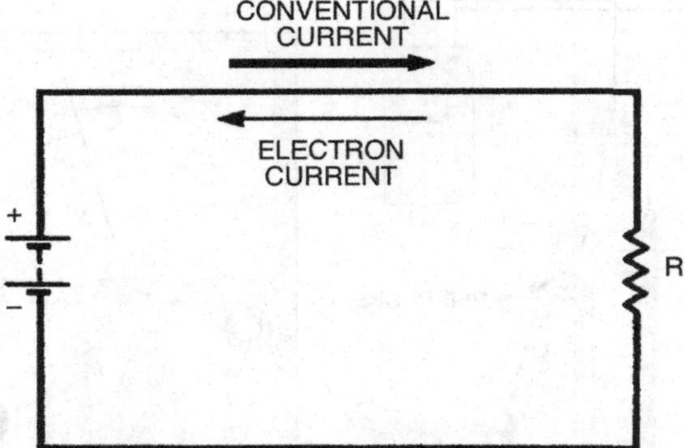

FIGURE 3.8 Conventional current travels from positive to negative; electron current travels in the opposite direction.

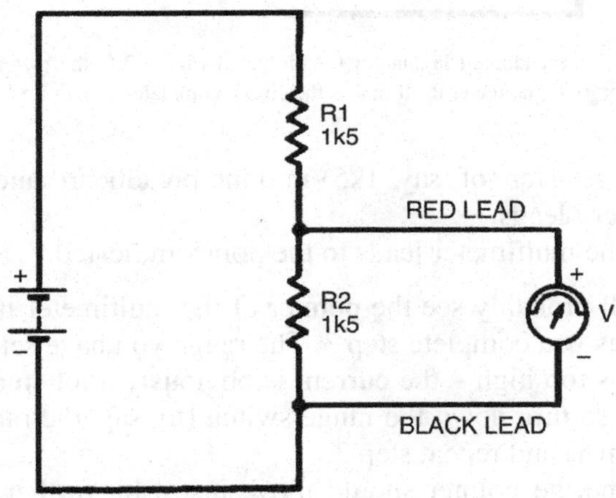

FIGURE 3.9 Meter connections to find the voltage across a single resistor.

PRACTICALLY THERE

Now that we've looked at the theory behind current and voltage measurement, let's go on and build a few circuits to put it into practice. Figure 3.10 shows the breadboard layout of the circuit of Figure 3.6, where a single resistor is connected in series with a multimeter and a battery. With this circuit we can actually prove Ohm's law. The procedure is as follows:

1. Set the meter's range switch to a current range – the highest one, say, 10 A.

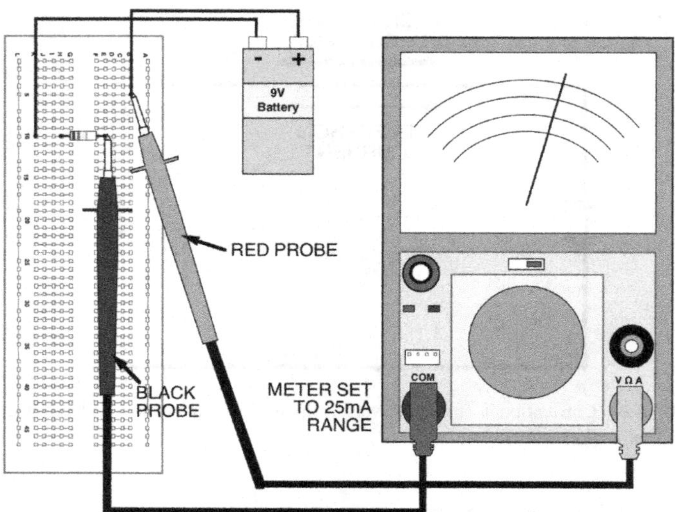

FIGURE 3.10 A breadboard layout for the circuit of Figure 3.6: the meter is in series with the resistor; only the meter itself makes the circuit complete.

2. Insert a resistor (of, say, 1k5) into the breadboard and connect the battery leads.
3. Touch the multimeter leads to the points indicated.

You will probably see the pointer of the multimeter move (but only just) as you complete step 3. The range you have set the multimeter to is too high – the current is obviously a lot smaller than this range, so turn down the range switch (to, say, the range nearest to 250 mA) and repeat step 3.

This time the pointer should move just a bit further, but still not enough to allow an accurate reading. Turn down the range switch again, this time to, say, the 25 mA range and repeat step 3. Now you should get an adequate reading, which should be 6 mA give or take small experimental errors.

Is this right? Let's compare it to the expression of Ohm's law:

$$I = \frac{V}{R}$$

If we insert known values of voltage and resistance into this we obtain:

$$I = \frac{9}{1500} = 6 \times 10^{-3}\ \text{A} = 6\ \text{mA}$$

Not bad, eh? We've proved Ohm's law!

You can try this again using different values of resistor if you wish, but don't use any resistors lower than about 500 Ω as you won't get an accurate reading, because the battery cannot supply currents of more than just a few milliamps. Even if it could, the resistor would overheat due to the current flowing through it.

In all cases where the battery is able to supply the current demanded by the resistor, the expression for Ohm's law will hold true.

Take Note

Make sure that you start a new measurement with the range switch set to the highest range and step down. This is a good practice to get into, because it may prevent damage to the multimeter by excessive currents. Even though circuits we look at in *Starting Electronics* will rarely have such high currents through them, there may be a time when you want to measure an unknown current or voltage in another circuit; if you don't start at the highest range – zap – your multimeter could be irreparably damaged.

Hint

The scale you must use to read any measurement depends on the range indicated by the meter's range switch. When the range switch points, say, to the 10 A range, the scale with the highest value of 10 should be used and any reading taken represents the current value. When the range switch points to, say, the 250 mA range, on the other hand (and also the other two ranges 2.5 and 25 mA), the scale with the highest value of 25 should be used. It all sounds tricky, doesn't it? Well, don't worry, once you've seen how to do it, you'll be taking measurements easily and quickly, just like a professional.

Note that current (and voltage) scales read in the opposite direction to the resistance scale we used in the last chapter, and they are linear. This makes them considerably easier to use than resistance scales and they are also more accurate, as you can more easily judge a value if the pointer falls between actual marks on the scale.

VOLTAGES

When you measure voltages with your multimeter the same procedure should be followed, using the highest voltage ranges first and stepping down as required. The voltages you are measuring here are all direct voltages as they are taken from a 9V d.c. battery. So you needn't bother using the three highest d.c. voltage ranges on the multimeter, as your 9V battery cannot generate a high enough voltage to damage the meter anyway. Also, don't bother using the a.c. voltage ranges as they are – pretty obviously – for measuring only alternating (that is, a.c.) voltages.

As an example you can build the circuit of Figure 3.9 up on your breadboard, as shown in Figure 3.11. What is the measured voltage? It should be about 4.5V.

Now measure the voltage across the other resistor – it's also about 4.5V. Well, that figures, doesn't it? There's about 4.5V across each resistor, so there is a total of 2×4.5V – that is, 9V – across them both: the voltage of the battery. This has demonstrated that resistors in series act as a voltage divider or a potential divider, dividing up the total voltage applied across them. It's understandable that the voltage across each resistor is the same and half the total voltage, because the two resistors are equal. But what happens if the two resistors aren't equal?

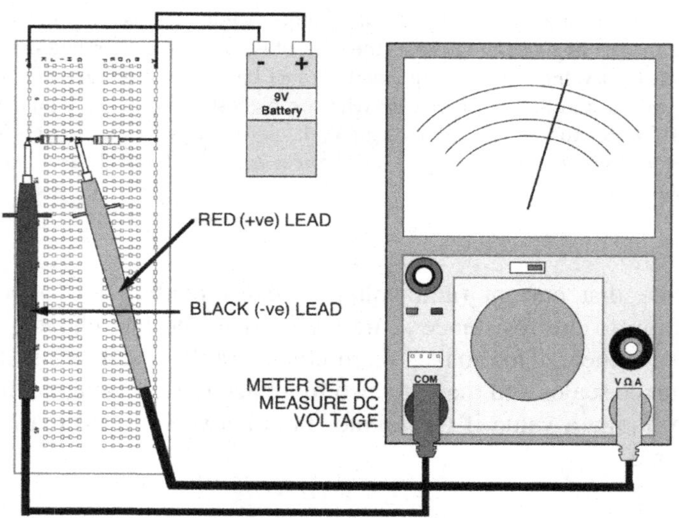

FIGURE 3.11 A breadboard layout for the circuit in Figure 3.9: measuring the voltage across R2. This will be the same as R1, and each will be around 4.5V – half the battery voltage.

Build up the circuit of Figure 3.12. What is the measured voltage across resistor R2 now? You should find it's about 2.1 V. The relationship between this result, the values of the two resistors, and the applied battery voltage is given by the voltage divider rule:

$$V_{out} = \frac{R2}{R1 + R2} \times V_{in}$$

where V_{in} is the battery voltage and V_{out} is the voltage measured across resistor R2.

We can check this by inserting the values used in the circuit of Figure 3.12:

$$\begin{aligned} V_{out} &= \frac{1k5}{4k7 + 1k5} \times 9 \\ &= \frac{13{,}500}{6200} \\ &= 2.18\ \text{V} \end{aligned}$$

In other words, close enough to our measured 2.1 V to make no difference. The voltage divider rule, like Ohm's law and the laws of series and parallel resistors, is one of the fundamental laws that we must know. So, remember it! OK?

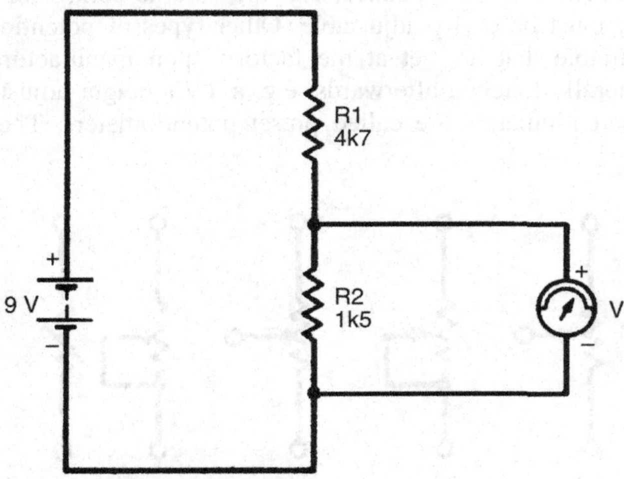

FIGURE 3.12 A circuit with two unequal series resistors. This is used in the text to illustrate the voltage divider rule, one of the most fundamental rules of electronics.

Take Note

By changing resistance values in a voltage divider, the voltage we obtain at the output is correspondingly changed. You can think of a voltage divider almost as a circuit itself, which allows an input voltage to be converted to a lower output voltage, simply by changing resistance values.

Pot-Heads

Certain types of components exist, ready-built for this voltage dividing job, known as potentiometers (commonly shortened to pots). They consist of some form of resistance track, across which a voltage is applied, and a wiper that can be moved along the track forming a variable voltage divider. The total resistance value of the potentiometer track doesn't change, only the ratio of the two resistances formed either side of the wiper. The basic symbol of a potentiometer is shown in Figure 3.13(a).

A potentiometer may be used as a variable resistor by connecting the wiper to one of the track ends, as shown in Figure 3.13(b). Varying the position of the wiper varies the effective resistance from zero to the maximum track resistance. This is useful if we wish to, say, control the current in a particular part of the circuit; increasing the resistance decreases the current and vice versa.

These two types of potentiometer are typically used when some function of an appliance, e.g. the volume control of a television, must be easily adjustable. Other types of potentiometer are available that are set at the factory upon manufacture and not generally touched afterwards, e.g. a TV's height adjustment. Such potentiometers are called preset potentiometers. The only

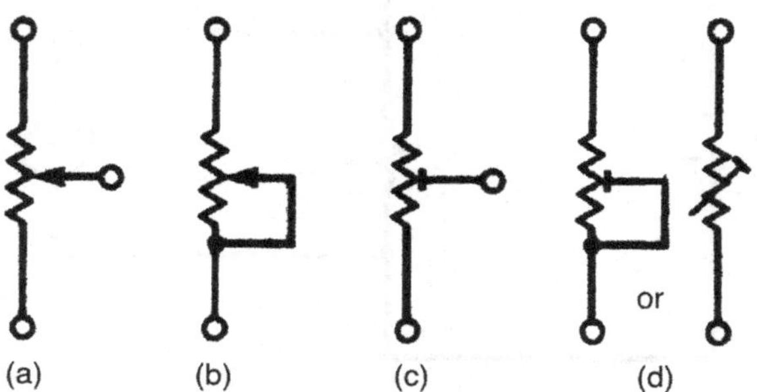

FIGURE 3.13 A variety of symbols used for variable resistors, or potentiometers.

difference as far as a circuit diagram is concerned is that their symbols are slightly changed. Figures 3.13(c) and (d) show preset potentiometers in the same configurations as the potentiometers of Figures 3.13(a) and (b). Mechanically, however, they are much different.

Meters Made

The actual internal resistance of a multimeter must be borne in mind when measuring voltages as it can affect measurements taken. We can build a circuit (Figure 3.14) which shows exactly what the effect of multimeter resistance is. As both resistors in the circuit are equal, we can see that the measured voltage should be half the battery voltage – that is, 4.5 V (use the voltage divider rule if you don't believe it!). But when you apply your multimeter across resistor R2 you find that the voltage indicated is only about 3 V.

The fact is that when the multimeter is not connected to the circuit the voltage is 4.5 V, but as soon as the multimeter is applied, the voltage across resistor R2 falls to 3 V. Also, the voltage across resistor R1 rises to 6 V (both voltages must add up to the battery voltage, remember). Applying the multimeter affects the operation of the circuit, because the multimeter resistance is in parallel with resistor R2.

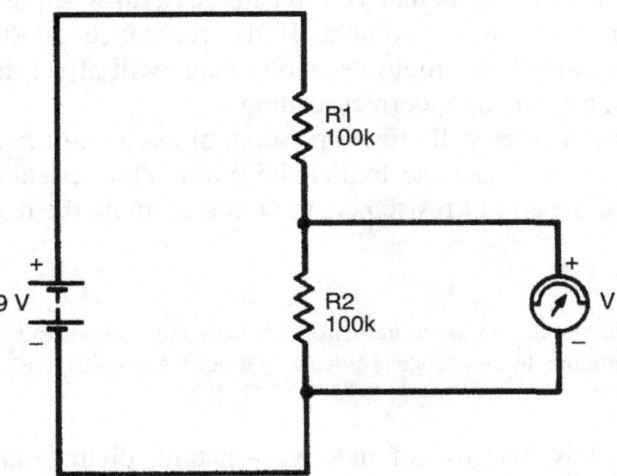

FIGURE 3.14 A circuit used to show the effect of the meter's own resistance in a circuit.

OHMS PER VOLT

A meter's internal resistance is stated as a number of ohms per volt. For example, the resistance of the meter we use in this book is 20,000 ohms per volt (written as $20\,\mathrm{k}\Omega\mathrm{V}^{-1}$) on d.c. voltage scales, so when it's to read 4.5 V, it's resistance is 90 k. This resistance, in parallel with resistor R2, forms an equivalent resistance, given by the law of parallel resistors, of:

$$R_{eq} = \frac{100\mathrm{k} \times 90\mathrm{k}}{100\mathrm{k} + 90\mathrm{k}} = 47\mathrm{k}4$$

This resistance is now the new value of R2 in the circuit so, applying the voltage divider rule, the measured voltage will be:

$$V = \frac{47\mathrm{k}4}{100\mathrm{k} + 47\mathrm{k}4} \times 9 = 2.9\ \mathrm{V}$$

which is roughly what you measured (hopefully).

Any difference between the actual measurement and this calculated value may be accounted for because this lower voltage causes a lower multimeter resistance, which in turn affects the voltage, which in turn affects the resistance, and so on until a balance is reached. All of this occurs instantly as soon as the multimeter is connected to the circuit.

What this tells us is that you must be careful when using your multimeter to measure voltage. If the resistance of the circuit under test is high, the multimeter resistance will affect the circuit operation, causing an incorrect reading.

Any multimeter will affect operation of any circuit to a greater or lesser extent – but the higher the multimeter resistance compared with the circuit resistance, the more accurate the reading.

Hint

A good rule-of-thumb to ensure reasonably accurate results is to make sure that the multimeter resistance is at least 10 times the circuit's resistance.

Fortunately, because of their very nature, digital multimeters usually have an internal resistance of the order of millions of ohms, a fact that allows it to accurately measure voltages in circuits with even high resistances.

Well. We've reached the end of this chapter, so how about giving the quiz that follows a go to see if you've really learned what you've been reading about.

QUIZ

Answers at the end of the book.

1. The voltage produced by a battery is:
   a. 9V
   b. Alternating
   c. 1.5V
   d. Direct
   e. None of these.
2. The voltage across the two resistors of a simple voltage divider:
   a. Is always 4.5V
   b. Is always equal
   c. Always adds up to 9V
   d. All of these
   e. None of these.
3. The voltage divider rule gives:
   a. The current from the battery
   b. The voltage across a resistor in a voltage divider
   c. The applied voltage
   d. b and c
   e. All of these
   f. None of these.
4. Conventional current flows from a point of higher electrical potential to one of lower electrical potential: true or false?
5. Applying a multimeter to a circuit, to measure one of the circuit's parameters, will always affect the circuit's operation to a greater or lesser extent: true or false?
6. Twenty-five, 100kΩ resistors are in parallel. What is the equivalent resistance of the network?:
   a. 4kΩ
   b. 2kΩ
   c. 2k0108
   d. 5kΩ
   e. None of these.

Chapter 4

Capacitors

You need a number of components to build the circuits in this chapter:

- 1 × 15k resistor
- 1 × 22 μF electrolytic capacitor
- 1 × 220 μF electrolytic capacitor
- 1 × 470 μF electrolytic capacitor
- 1 × switch.

As usual, the resistor power rating and tolerance are of no importance. The capacitor must, however, have a voltage rating of 10V or more. The switch can be of any type, even an old light switch will do if you have one. A slide switch is ideal, however, both common and cheap.

To make the connections between the switch and your breadboard, you can use common multi-strand wire if you like, and tin the ends that push into the breadboard – see Chapter 3 for instructions – but a better idea is to use 22 SWG single-strand tinned copper wire. This is quite rigid and pushes easily into the breadboard connectors without breaking up. It's available in reels, and a reel will last you quite a while, so it's worth buying for this and future purposes.

Take Note

If short strands break off from multi-strand wire and get stuck in the breadboard connectors you might get short circuits occurring. Make sure – if you use multi-strand wire – you tin wire ends properly. Even better – don't use multi-strand wire, instead use single-strand wire.

Solder a couple of short lengths of the wire to switch connections and push on a couple of bits of insulating covering (you can buy this in reels also, but if you have a few inches of spare mains

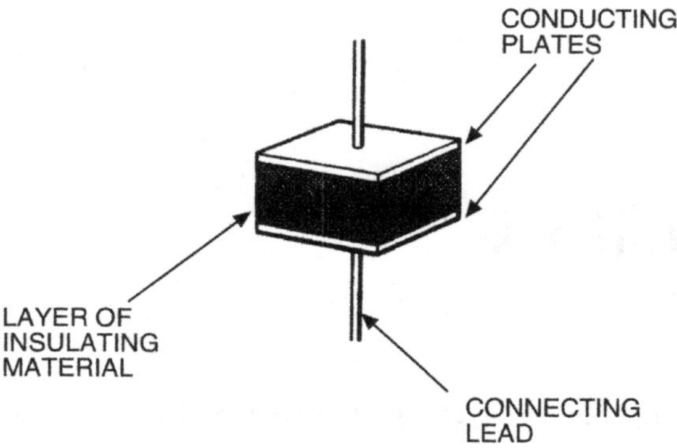

FIGURE 4.1 A simple diagrammatic representation of a capacitor.

cable you'll find you can strip the few centimeters you need from this) to protect against short circuits.

CAPACITORS

We're going to take a close look at a new type of component this chapter: the capacitor. We will see how capacitors function, firstly in practical experiments that you can do for yourself and then, secondly, with a bit of theory (boring, boring, boring! Who said that?), which will show you a few things that can't easily be demonstrated in experiments.

So, let's get started. What exactly makes up a capacitor? Well, there's an easy answer: a capacitor is made up from two electrically conductive plates, separated from each other by a sandwiched layer of insulating material. Figure 4.1 gives you a rough idea.

However, this is an oversimplification really; there are many, many different types of capacitor and, although their basic make-up is of two insulated plates, they are not always made in quite this simple form. Nevertheless, the concept is one to be going along with.

DOWN TO BUSINESS

Figure 4.2(a) shows the circuit symbol of an ordinary capacitor. As you'll appreciate, this illustrates the basic capacitor make-up we've just seen. Figure 4.2(b), on the other hand, shows the circuit symbol of a specific, although very common, type of capacitor: the electrolytic capacitor.

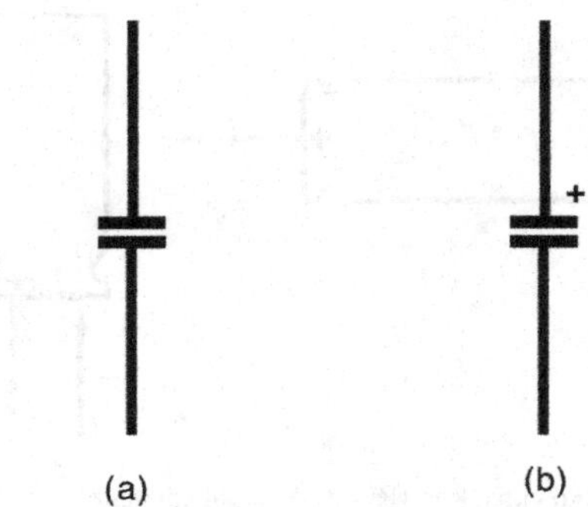

FIGURE 4.2 Circuit symbols for an ordinary (a) and an electrolytic (b) capacitor.

Hint

The electrolytic capacitor gets its name from the fact that its basic capacitor action is derived through the process of electrolysis when the capacitor is connected into a circuit. This electrolytic action means that an electrolytic capacitor must be inserted into the circuit the right way round, unlike most other capacitors. We say that electrolytic capacitors are polarized and to show this the circuit symbol has positive (white) and negative (black) plates. Sometimes, as we've shown, a small positive symbol (+) is drawn beside the electrolytic capacitor's positive plate to reinforce this.

In our experiments in this chapter, we're going to use electrolytic capacitors, not because we like to be awkward, but because the values of capacitance we want are quite high. And electrolytic capacitors are generally the only ones capable of having these values, while keeping to a reasonable size and without being too expensive. Anyway, not to worry, all you have to do is remember to put the capacitors into the circuit the right way round.

You can make sure of this, as all electrolytic capacitors have some kind of marking on them that identifies positive and negative plates. Figure 4.3 shows two types of fairly typical electrolytic capacitors. One, on the left, has what is called an axial body, where the connecting leads come out from each end. One end, the positive plate end, generally has a ridge around it and sometimes is marked with positive symbols (as mine is). Sometimes, the negative plate end has a black band around it, or negative symbols.

The capacitor on the right has a radial body, where both leads come out from one end. Again, however, one or sometimes both of the leads will be identified by polarity markings on the body.

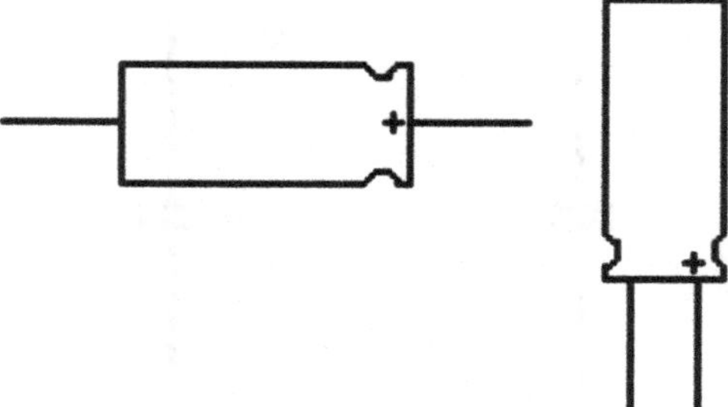

FIGURE 4.3 (Left) An axial capacitor. (Right) A radial capacitor.

Whatever type of electrolytic capacitor we actually use in circuits, we will show the capacitor in a breadboard layout diagram as an axial type. This is purely to make it obvious (due to the ridged positive plate end) which lead is which. Both types are, in fact, interchangeable as long as the correct polarity is observed, and voltage rating (that is, the maximum voltage that can be safely applied across its leads – usually written on an electrolytic capacitor's body) isn't exceeded.

MEASURING UP

Capacitance values are measured in a unit called the farad (named after the scientist Faraday) and given the symbol: F. We'll define exactly what the farad is later; suffice to say here that it is a very large unit. Typical capacitors have values much, much smaller. Fractions such as a millionth of a farad (that is, one microfarad: 1 μF), a thousand millionth of a farad (that is, one nanofarad: 1 nF), or one million millionth of a farad (that is, one picofarad: 1 pF) are common. Sometimes, like the Ω or R of resistances, the unit F is omitted – 15 n instead of 15 nF, say.

The circuit for our first experiment is shown in Figure 4.4. You will see that it bears a striking resemblance to the voltage divider resistor circuits we looked at in the last chapter, except that a capacitor has taken the place of one of the resistors. Also included in the circuit of Figure 4.4 is a switch. In the last chapter's circuits, no switch was used as we were only interested in static conditions after the circuits were connected. The circuits in this chapter, on the other hand, have a number of dynamic properties that occur for only a while after the circuit is connected. We can use the switch to make sure that we see all of these dynamic

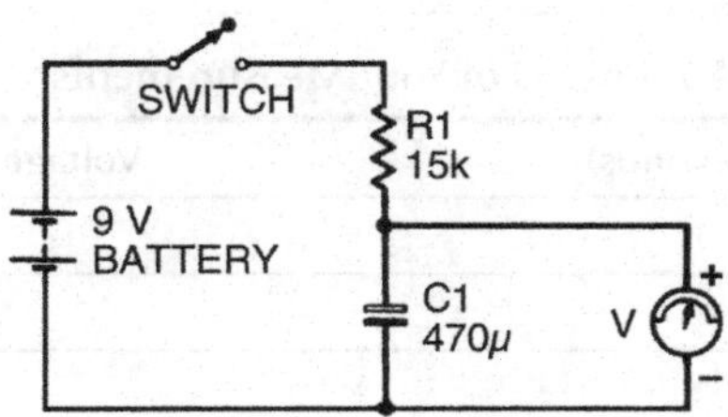

FIGURE 4.4 A simple resistor and capacitor circuit.

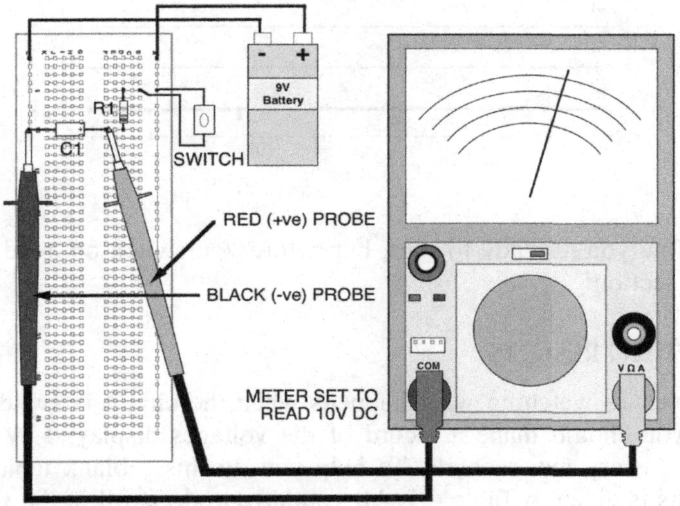

FIGURE 4.5 A breadboard layout for the circuit in Figure 4.4.

properties, from the moment electric current is allowed to flow into the circuit.

A breadboard layout for the circuit is shown in Figure 4.5. Make sure the switch is off before you connect your battery up to the rest of the circuit.

As the switch is off, no current should flow so the meter reading should be 0 V. If you've accidentally had the switch on and current has already flowed, you'll have a voltage reading other than zero, even after the switch is consequently turned off. In such a case, get a short length of wire and touch it to both leads of the capacitor simultaneously, for a few seconds, so that a short circuit is formed. The voltage displayed by the meter should rapidly fall to 0 V and stay there.

TABLE 4.1 Results of Your Measurements

Time (seconds)	Voltage (V)
5	
10	
15	
20	
30	
40	
50	
60	

Now you're ready to start, but before you switch on, read the next section!

GETTING RESULTS

As well as watching what happens when the circuit is switched on, you should make a record of the voltages displayed by the meter, every few seconds. To help you do this a blank table of results is given in Table 4.1. All you have to do is fill in the voltages you have measured, at the times given, into the table at successive measurement points. You'll find that the voltage increases from zero as you switch the circuit on, rapidly at first but slowing down to a snail's pace at the end. Don't worry if your measurements of time and voltage aren't too accurate – we're only trying to prove the principle of the experiment, not the exact details. Besides, if you switch off and then short-circuit the capacitor, as already described, you can repeat the measurements as often as required.

When you've got your results, transfer them as points on to the blank graph of Figure 4.6 and sketch in a curve that goes approximately through all the points. Table 4.2 and Figure 4.7 show the results obtained when the experiment was performed in preparing this book.

What you should note in your experiment is that the voltage across the capacitor (any capacitor, in fact) rises, not instantaneously

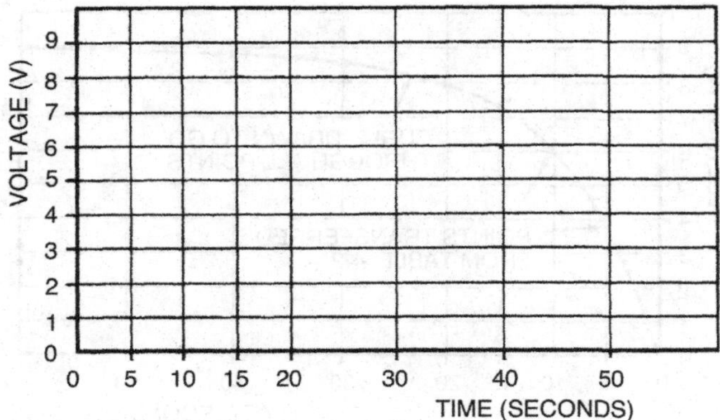

FIGURE 4.6 A blank graph to plot your measurements. Use Table 4.1 with it to record your measurements.

TABLE 4.2 Our Results While Preparing This Book

Time (seconds)	Voltage (V)
5	4
10	6
15	7
20	8
30	8.5
40	8.7
50	8.9
60	9

as with the voltage across a resistor, but gradually, following a curve. Is the curve the same, do you think, for all capacitors? Change the capacitor for one with a value of 220 μF (about half the previous one). Using the blank table of Table 4.3 and blank graph in Figure 4.8, perform the experiment again to find out.

Table 4.4 and Figure 4.9 show the results of our second experiment (yours should be similar); it is seen that although the curves of the two circuits aren't exactly the same as far as the time axis is concerned, they are exactly the same shape. Interesting, huh?

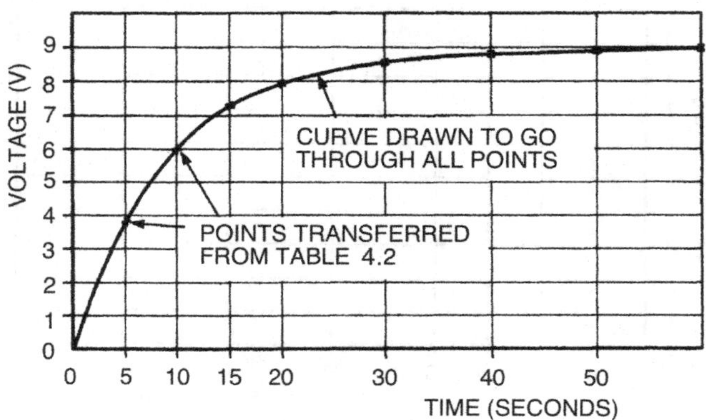

FIGURE 4.7 The results of our experiments.

TABLE 4.3 The Results of Your Measurements

Time (seconds)	Voltage (V)
5	
10	
15	
20	
25	
30	
40	

This proves that the rising voltage across a capacitor follows some sort of law. It is an exponential law, and the curves you obtained are known as exponential curves. These exponential curves are related by their time constants. We can calculate any rising exponential curve's time constant, as shown in Figure 4.10, where a value of 0.63 times the total voltage change cuts the time axis at a time equaling the curve's time constant, τ (the Greek letter tau).

In a capacitor circuit like that of Figure 4.4, the exponential curve's time constant is given simply by the product of the

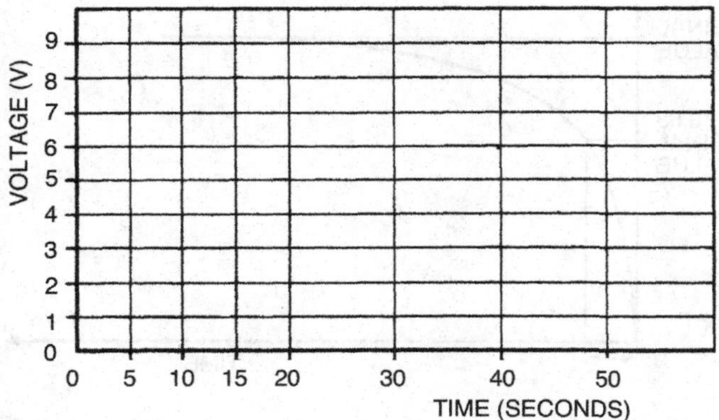

FIGURE 4.8 A blank graph to plot your measurements. Use Table 4.3 as well.

TABLE 4.4 The Results of Our Second Experiment

Time (seconds)	Voltage (V)
5	6
10	7.2
15	8
20	8.3
25	8.5
30	8.7
40	9

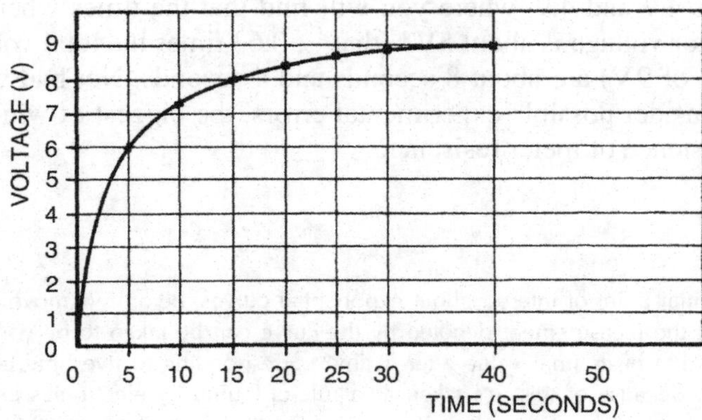

FIGURE 4.9 A graph showing our results of the second experiment.

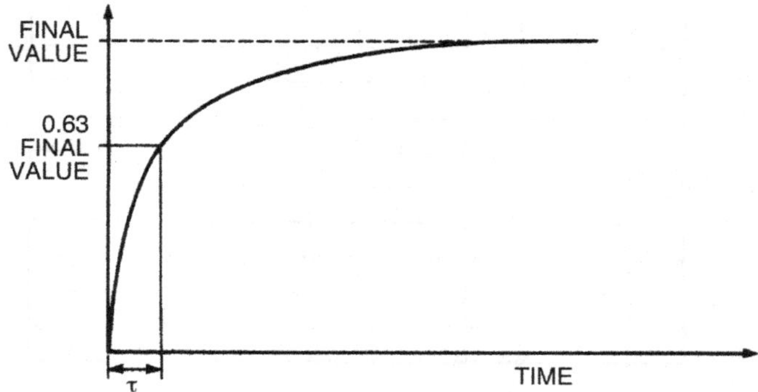

FIGURE 4.10 A graph showing an exponential curve, related to the time constant.

capacitor and resistor value. So, in the case of the first experiment, the time constant is:

$$\begin{aligned}\tau_1 &= 470 \times 10^{-6} \times 15{,}000 \\ &= 7 \text{ seconds}\end{aligned}$$

and, in the case of the second experiment:

$$\begin{aligned}\tau_2 &= 220 \times 10^{-6} \times 15{,}000 \\ &= 3.3 \text{seconds}\end{aligned}$$

You can check this, if you like, against your curves, or those in Figures 4.7 and 4.9, where you will find that the times when the capacitor voltage is about 5V7 (that is, 0.63 times the total voltage change of 9V) are about 8 seconds and 4 seconds. Not bad when you consider possible experimental errors, the biggest of which is the existence of meter resistance.

Hint

One final point of interest about exponential curves, before we move on, is that the measurement denoted by the curve can be taken to be within about 1% of its final value after a time corresponding to five time constants. Because of this, it's taken as a rule-of-thumb by electronics engineers that the changing voltage across a capacitor is complete after 5τ.

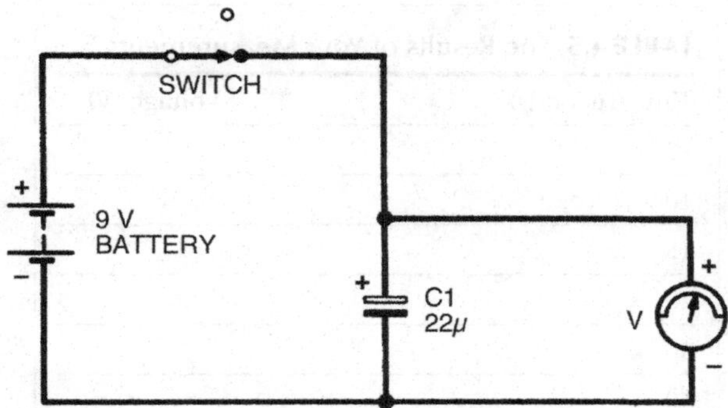

FIGURE 4.11 An experimental circuit to demonstrate how a capacitor stores voltage (or charge).

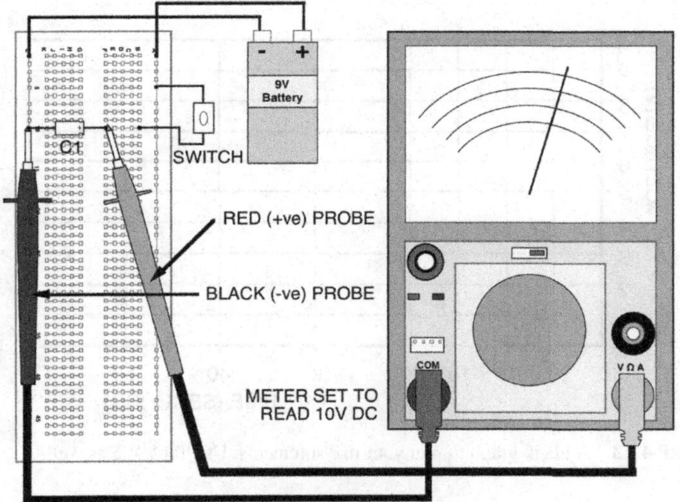

FIGURE 4.12 The breadboard layout for the experiment shown in Figure 4.11.

THE OTHER WAY

During these experiments you should have noticed how the voltage across the capacitor appeared to stay for a while, even after the switch had been turned off. How could this be so?

Figure 4.11 shows the circuit of an experiment you should now do, which will help you to understand this unusual capability of capacitors to apparently store voltage. Build the circuit up as in the breadboard layout of Figure 4.12, with the switch in its on position.

TABLE 4.5 The Results of Your Measurements

Time (seconds)	Voltage (V)
5	
10	
15	
20	
30	
40	
50	
60	

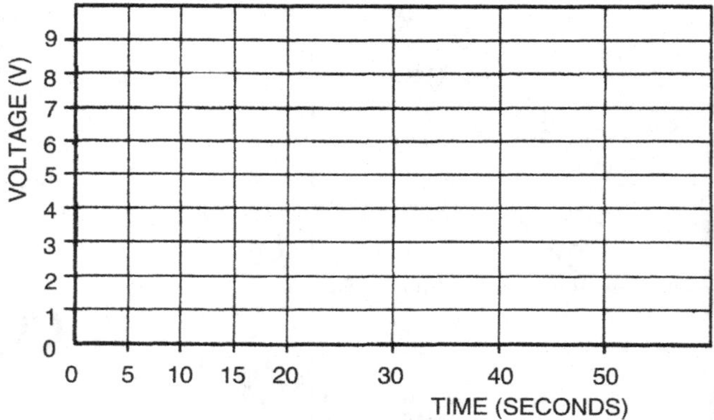

FIGURE 4.13 A blank graph to plot your measurements. Use Table 4.5 as well.

Now, measure the voltages at selected time intervals, and fill in the results into Table 4.5, then transfer them to form a curve in Figure 4.13, after switching the switch off.

The results we obtained in preparing this book are tabulated in Table 4.6 and plotted in Figure 4.14. You can see that the curve obtained is similar to the curves we got earlier, but are falling not rising. Like the earlier curves, this is an exponential curve too.

If you wish, you can calculate the time constant here in the same way, but remember that it must now be the time at which the voltage falls to 0.63 of the total voltage drop – that is, when:

$$V = 9 - (0.63 \times 9)$$
$$= 3.3 \text{ V}$$

TABLE 4.6 The Results of Our Experiments While Preparing This Book

Time (seconds)	Voltage (V)
5	7.5
10	6
15	5
20	4
30	2.5
40	1.5
50	1
60	0.7

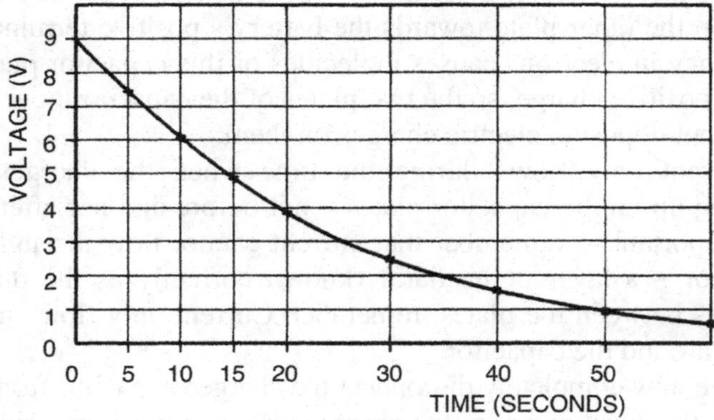

FIGURE 4.14 The graph showing our own results.

THEORETICAL ASPECTS

So what is it about a capacitor that causes this delay in voltage between switching on or off the electricity supply and obtaining the final voltage across it? It's almost as if there's some mystical time delay inside the capacitor. To find the answer we'll have to consider a capacitor's innards again.

Figure 4.15 shows the capacitor we first saw in Figure 4.1, but this time it is shown connected to a battery that, as we know, is capable of supplying electrons from its negative terminal.

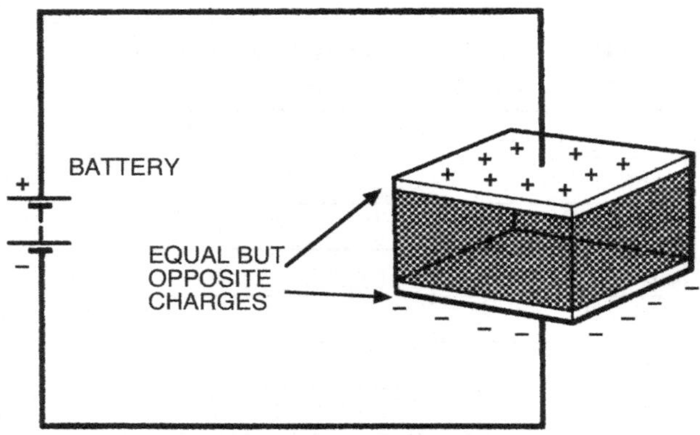

FIGURE 4.15 The capacitor shown in Figure 4.1, but this time connected to a battery with opposite charges built up on each plate.

A number of electrons gather on the capacitor plate connected to the battery's negative terminal, which in turn repel any electrons on the other plate towards the battery's positive terminal. A deficiency in electrons causes molecules of this capacitor plate to have a positive charge, so the two plates of the capacitor now have equal, but opposite, electric charges on them.

Current only flows during the time when the charges are building up on the capacitor plates – not before and not after. It's also important to remember that current cannot flow through the capacitor – a layer of insulator (known correctly as the dielectric) lies between the plates, remember. Current only flows in the circuit around the capacitor.

If we now completely disconnect the charged capacitor from the battery, the equal and opposite charges remain – in theory – indefinitely. In practice, on the other hand, charge is always lost due to leakage current between the plates. You can try this for yourself, if you like: put a capacitor into the breadboard then charge it up by connecting the battery directly across it. Now disconnect the battery and leave the capacitor in the breadboard for a time (overnight, say). Then connect your meter across it to measure the voltage. You should still get a reading, but remember that the resistance of the meter itself will always drain the charge stored.

The size of the charge stored in a capacitor depends on two factors: the capacitor's capacitance (in farads) and the applied voltage. The relationship is given by:

$$Q = C \times V$$

where Q is the charge measured in coulombs, C is the capacitance, and V is the voltage. From this we can see that a charge of 1 coulomb is stored by a capacitor of 1 farad, when a voltage of 1 volt is applied.

Alternatively, we may define the farad (as we promised we would, earlier in the chapter) as being the capacitance that will store a charge of 1 coulomb when a voltage of 1 volt is applied.

We can now understand why it is that changing the capacitor value changes the time constant, and hence changes the associated time delay in the changing voltage across the capacitor. Increasing the capacitance, say, increases the charge stored. As the current flowing is determined by the resistance in the circuit, and is thus fixed at any particular voltage, this increased charge takes longer to build up or longer to decay away. Reducing the capacitance reduces the charge, which is therefore more quickly stored or more quickly discharged.

Similarly, as the resistor in the circuit defines the current flowing to charge or discharge the capacitor, increasing or decreasing its value must decrease or increase the current, therefore increasing or decreasing the time taken to charge or discharge the capacitor. This is why the circuit's time constant is a product of both capacitance and resistance values.

CAPACITANCE VALUES

Finally, we can look at how the size and construction of capacitors affects their capacitance. The capacitance of a basic capacitor, for example, consisting of two parallel, flat plates separated by a dielectric, may be calculated from the formula:

$$C = \varepsilon \times \frac{A}{d}$$

where ε is the permittivity of the dielectric, A is the area of overlap of the plates, and d is the distance between the plates.

Different insulating materials have different permittivities, so different capacitor values may be made, simply by using different dielectrics. Air, say, has a permittivity of about 9×10^{-12} farads per meter (that is, $9 \times 10^{-12}\ \mathrm{F\,m^{-1}}$). Permittivities of other materials are usually expressed as a relative permittivity, where a material's relative permittivity is the number of times its permittivity is greater than air. So, for example, mica (a typical capacitor dielectric), which has a relative permittivity of 6, has an actual permittivity of $54 \times 10^{-12}\,\mathrm{F\,m^{-1}}$ (that is, $6 \times 9 \times 10^{-12}\,\mathrm{F\,m^{-1}}$).

So, a possible capacitor, consisting of two parallel metal plates of overlapping area $20\,\mathrm{cm}^2$, 10 mm apart, with a mica dielectric, has a capacitance of:

$$\begin{aligned} C &= 54 \times 10^{-12} \times \frac{0.002}{0.01} \\ &= 10.8 \times 10^{-12} \\ &= 10.8\,\mathrm{pF} \end{aligned}$$

Er, I think that's enough theory to be grappling with for now, don't you? Try your understanding of capacitors and how they work with the chapter's quiz.

QUIZ

Answers at the end of the book.

1. The value of a capacitor with two plates of overlapping area $40\,\mathrm{cm}^2$, separated by a 5 mm layer of air, is:
   a. 7.2 pF
   b. 7.2×10^{-12} pF
   c. 72 pF
   d. 100 μF
   e. a and b
   f. None of these.
2. A capacitor of 1 nF has a voltage of 10 V applied across it. The maximum charge stored is:
   a. 1×10^{-10} C
   b. 1×10^{8} C
   c. 10×10^{-8} C
   d. 1×10^{-8} F
   e. a and c
   f. None of these.
3. A capacitor of 10 μF and a resistor of 1 MΩ are combined in a capacitor/resistor charging circuit. The time constant is:
   a. 1 second
   b. 10×10^{6} seconds
   c. 10 seconds

d. Not possible to calculate from these figures
e. None of these.

4. A capacitor is fully charged by an applied voltage of 100 V, then discharged. After a time not less than 40 seconds the voltage across the capacitor has fallen to 1 V. At what time was the voltage across the capacitor 37 V?:
a. 10 seconds
b. 37 seconds
c. 8 seconds
d. It is impossible to say from the figures given
e. None of these.

5. In question 4, the capacitor has a value of 100 μF. What is the charging resistance value?:
a. 80,000 μF
b. 80,000 C
c. 80 kΩ
d. 8 kΩ
e. A slap in the face with a wet fish
f. None of these.

Chapter 5

ICs, Oscillators, and Filters

You'll need a number of different components to build the circuits in this chapter; some of them you'll already have, but there's a few new ones too. First, the resistors required are:

- 1 × 1k5
- 1 × 4k7
- 1 × 10k.

Second, the capacitors needed are:

- 1 × 1 nF
- 2 × 10 nF
- 2 × 100 nF
- 1 × 1 µF electrolytic
- 1 × 10 µF electrolytic.

The power ratings, tolerances, and so on of all these components are not critical, but the electrolytic capacitors should have a voltage rating of 9 V or more.

Some other components you already have – a switch, battery, battery connector, breadboard, and multimeter should all be close at hand, as well as some single-strand tinned copper wire. You are going to use the single-strand wire to make connections from point to point on the breadboard, and as it's uninsulated this has to be done carefully, to prevent short circuits. Photo 5.1 shows the best method. Cut a short length of wire and hold it in the jaws of your snipe-nosed or long-nosed pliers. Bend the wire round the jaws to form a sharp right-angle in the wire. The tricky bit is next – judging the length of the connection you require, move the pliers along the wire and then bend the other end of the wire also at right angles round the other side of the jaws (Photo 5.2). If you remember that the holes in the grid of the breadboard are equidistantly spaced at 2.5 mm (or a tenth of an inch if you're old-fashioned like me – what's an inch, Grandad?), then it becomes easier.

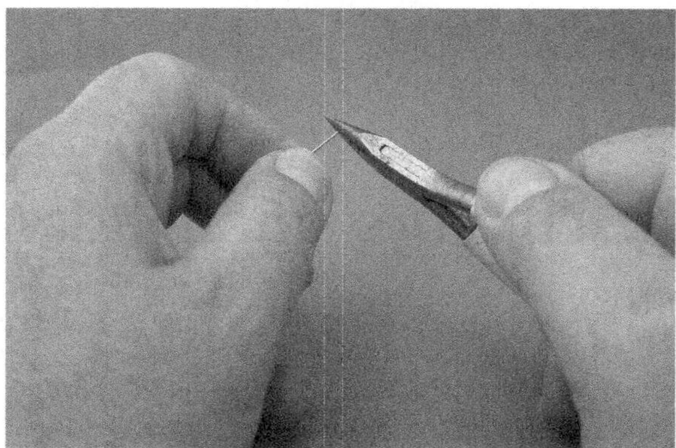

PHOTO 5.1 Take a short piece of wire between the jaws of your pliers ….

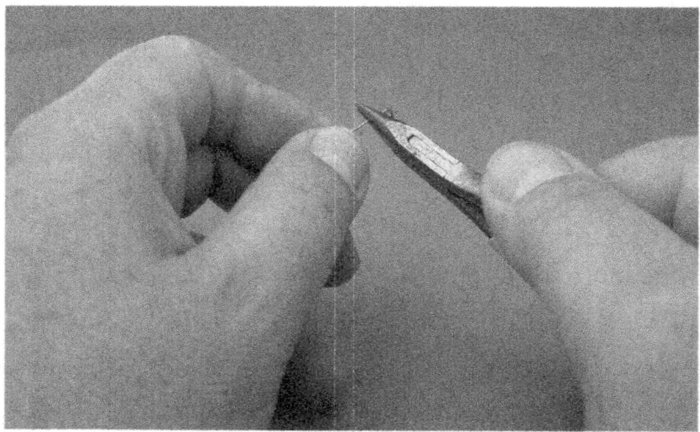

PHOTO 5.2 … and bend it carefully into two right angles, judging the length.

A connection over two holes is 5 mm long, over four holes 10 mm long, and so on – you'll soon get the hang of it.

Now, holding the wire at the top, in the pliers, push it into the breadboard as shown in Photo 5.3, until it lies flush on the surface of the breadboard as in Photo 5.4. No bother, eh? Even with a number of components in the breadboard it is difficult to short-circuit connections made this way.

Two other components you need are:

- 1 × 555 integrated circuit
- 1 × light-emitting diode (any color).

We've seen an integrated circuit (IC) before and we know what it looks like, but we've never used one before, so we'll take

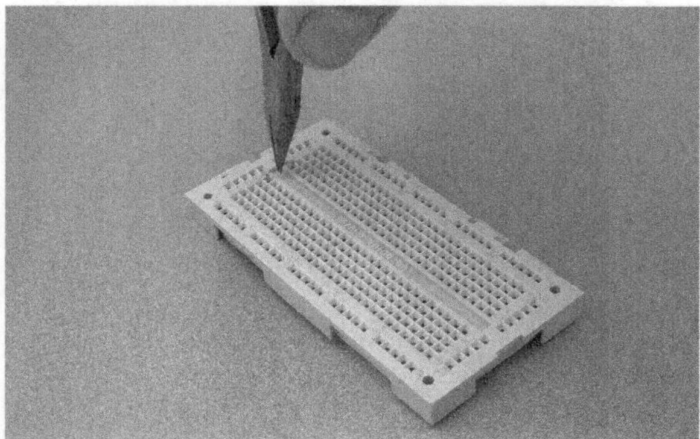

PHOTO 5.3 The wire link you have made should drop neatly into the breadboard.

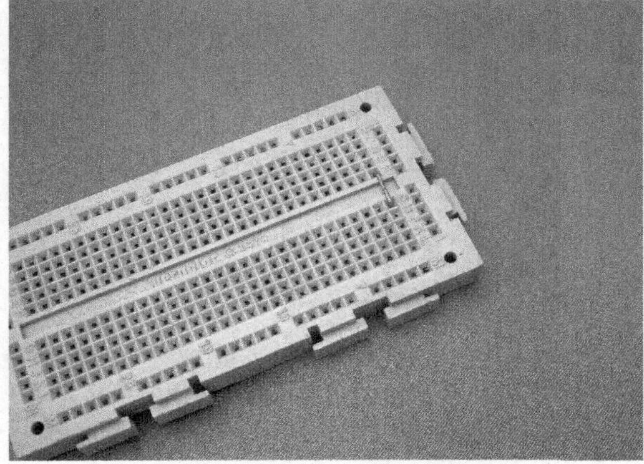

PHOTO 5.4 If the link is a good fit, it can be used over and over again in different positions.

a closer look at the 555 now. It is an eight-pin dual-in-line (DIL) device and one is shown in Photo 5.5. Somewhere on its body is a notch or dot, which indicates the whereabouts of pin 1 of the IC, as shown in Figure 5.1. The remainder of the pins are numbered in sequence in an anticlockwise direction around the IC.

ICs should be inserted into a breadboard across the breadboard's central bridged portion. Isn't it amazing that this portion is 7.5 mm across and, hey presto, the rows of pins of the IC are about 7.5.mm apart? It's as if the IC was made for the breadboard! So the IC fits into the breadboard something like that shown in Photo 5.6.

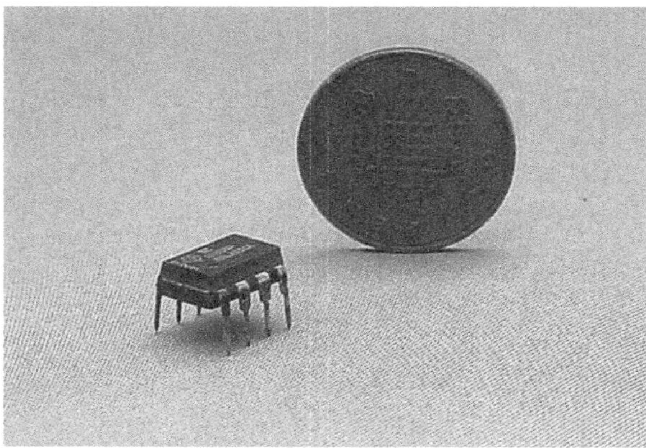

PHOTO 5.5 The 555, an eight-pin DIL timer IC, beside a UK one penny piece.

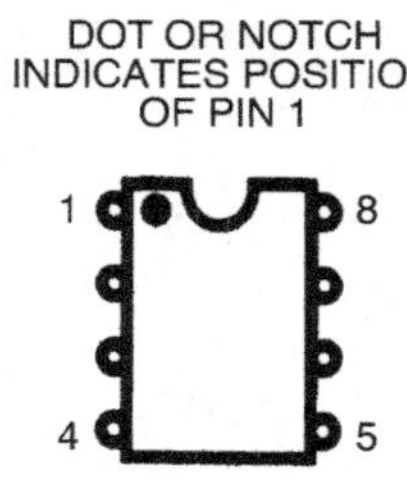

FIGURE 5.1 A diagram of the normal configuration of any IC: the dot marks pin 1 and the remaining pins run anticlockwise.

PHOTO 5.6 The IC mounted across the central divide in the breadboard, which is designed to be the exact size.

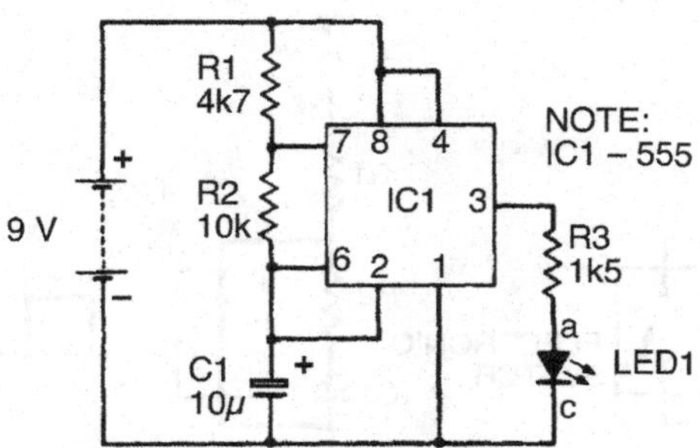

FIGURE 5.2 The circuit of the astable multi-vibrator circuit using the 555.

We'll look at the light-emitting diode later.

If you remember, in the last chapter we took a close look at capacitors, how they charge and discharge, storing and releasing electrical energy. The first thing we'll do in this chapter, however, is use this principle to build a useful circuit called an oscillator. Then, in turn, we'll use the oscillator to show some more principles of capacitors. So, we've got a twofold job to do now and there's an awful lot of work to get through – let's get started.

Figure 5.2 shows the circuit of the oscillator we're going to build. It's a common type of oscillator known as an astable multivibrator. The name arises because the output signal appears to oscillate (or vibrate) between two voltages, never resting at one voltage for more than just a short period of time (it is therefore unstable, more commonly known as astable). An astable multivibrator built from discrete – that is, individual – components can be tricky to construct, so we've opted to use an integrated circuit (the 555) as the oscillator's heart.

Inside the 555 is an electronic switch that turns on when the voltage across it is approximately two-thirds the power supply voltage (about 6 V in the circuit of Figure 5.2), and off when the voltage is less than one-third the power supply voltage (about 3 V). Figure 5.3(a) shows an equivalent circuit to that of Figure 5.2 for the times during which the electronic switch of the 555 is off.

You should be able to work out that the capacitor C1 of the circuit is connected through resistors R1 and R2 to the positive power supply rail. The time constant τ_1 of this part of the circuit is therefore given by:

$$\tau_1 = RC = (\text{R1} + \text{R2})C$$

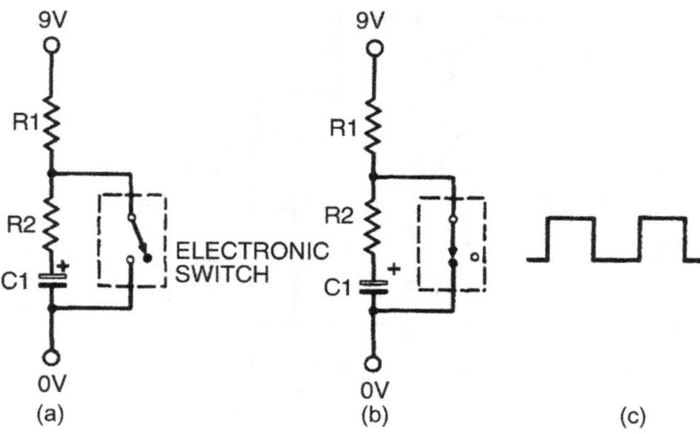

FIGURE 5.3 (a) A circuit equivalent to Figure 5.2 when the 555's switch is off. (b) The circuit when the switch is on. (c) The square wave output signal.

When the voltage across the switch rises to about 6 V, however, the switch turns on (as shown in Figure 5.3(b)), forming a short circuit across the capacitor and resistor R2. The capacitor now discharges with a time constant given by:

$$\tau_1 = \mathrm{R2} \times C$$

Of course, when the discharging voltage across the switch falls to about 3 V, the switch turns off again, and the capacitor charges up once more. This process repeats indefinitely, with the switch turning on and off at a rate determined by the two time constants. Because of this up and down effect, such oscillators are often known as relaxation oscillators.

As you might expect, the circuit integrated within the 555 is not that simple and there are many other parts to it (one part, for example, converts the charging and discharging exponential voltages into only two definite voltages – 9 and 0 V – so that the 555's output signal is a square wave, as shown in Figure 5.3(c)). But the basic idea of the astable multi-vibrator formed by a 555 is just as we've described here.

THROWING LIGHT ON IT

The 555 IC is one of only two new types of electronic component this circuit introduces you to. The other is a light-emitting diode (LED), which is a type of indicator. One is shown in Photo 5.7. LEDs are polarized and so must be inserted into a circuit the

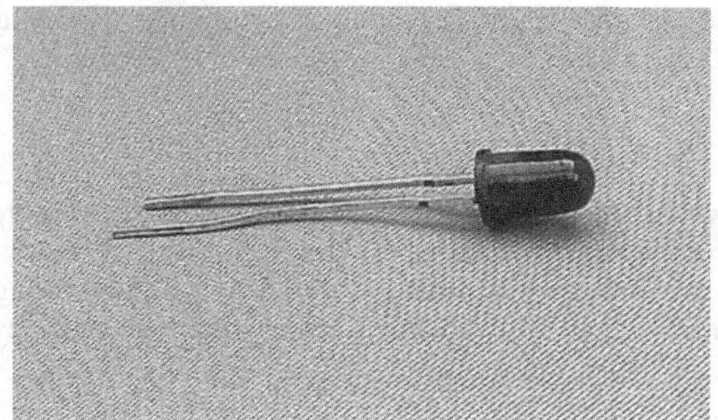

PHOTO 5.7 One of the most popular electronic components, the LED.

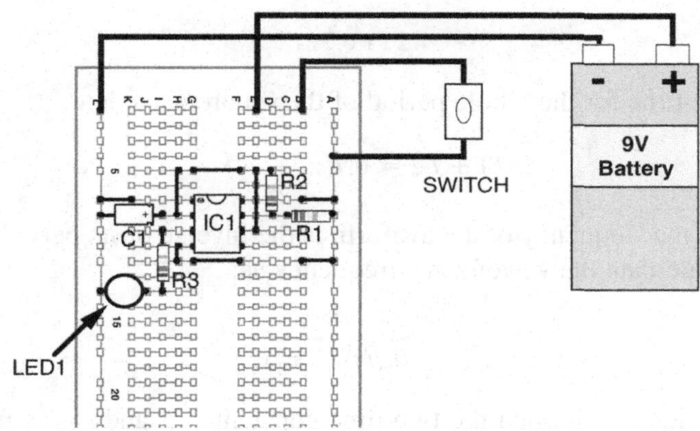

FIGURE 5.4 The breadboard layout for the multi-vibrator circuit as shown in Figure 5.2.

right way round. All LEDs have an anode (which goes to the more positive side of the circuit) and a cathode (which goes to the more negative side).

Generally, but not always, the anode and cathode of an LED are identified by the lengths of the component leads – the cathode is the shorter of the two. There's often a flat side to an LED too – usually, but again not always, on the cathode side – to help identify leads.

The complete circuit's breadboard layout is shown in Figure 5.4. Build it and see what happens.

When you turn on, you should find that the LED flashes on and off quite rapidly (about five or six times a second, actually). This means your circuit is working correctly. If it doesn't work, check the polarity of all polarized components: the battery, IC, LED, and capacitor.

OUCH, THAT HERTZ

We can calculate the rate at which the LED flashes, more accurately, from formulae relating to the 555. A quick study of the square wave output shows that it consists of a higher voltage for a time (which we can call $T1$) and a lower voltage for a time (which we will call $T2$).

Now, $T1$ is given by:

$$T1 = 0.7\tau_1$$

and $T2$ is given by:

$$T2 = 0.7\tau_2$$

So, the time for the whole period of the square wave is:

$$T1 + T2 = 0.7(\tau_1 + \tau_2)$$

and as the frequency of a waveform is the inverse of its period, we may calculate the waveform's frequency as:

$$f = \frac{1}{0.7(\tau_1 + \tau_2)}$$

Earlier, we defined the two time constants, τ_1 and τ_2, as functions of the capacitor and the two resistors, and so by substituting them into the above formula, we can calculate the frequency as:

$$f = \frac{1}{0.7C(\mathrm{R1} + 2\mathrm{R2})} \quad (5.1)$$

So, the frequency of the output signal of the circuit of Figure 5.2 is:

$$\begin{aligned} f &= \frac{1}{0.7 \times 10 \times 10^{-6} \times (4700 + 20{,}000)} \\ &= 5.8 \text{ cycles per second} \end{aligned}$$

or, more correctly speaking:

$$= 5.8 \text{ hertz (shortened to 5.8 Hz)}$$

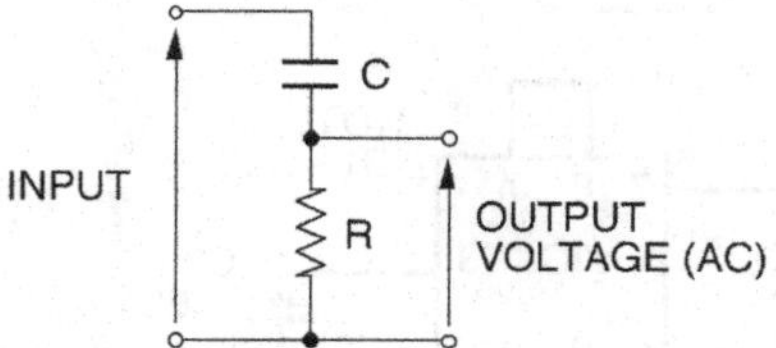

FIGURE 5.5 A voltage divider – but with a capacitor.

Equation (5.1) is quite important really, because it shows that the frequency of the signal is inversely proportional to the capacitance. If we decrease the value of the capacitor we will increase the frequency. We can test this by taking out the 10 μF capacitor and putting in a 1 μF capacitor. Now, the LED flashes so quickly (about 58 times a second) that your eye cannot even detect it is flashing and it appears to be always on. If you replace the capacitor with one of a value of, say, 100 μF the LED will flash only very slowly.

Now, let's stop and think about what we've just done. Basically we've used a capacitor in precisely the ways we looked at in the last chapter – to charge and discharge with electrical energy so that the voltage across the capacitor goes up and down at the same time. True, in the experiments in the last chapter you were the switch, whereas in this chapter an IC has taken your place. But the principle – charging and discharging a capacitor – is the same.

The current that enters the capacitor to charge it, then leaves the capacitor to discharge it, is direct current because it comes from a 9 V d.c. battery. However, if we look at the output signal (Figure 5.3(c)) we can see that the signal alternates between two levels. Looked at in this way, the astable multi-vibrator is a d.c.-to-a.c. converter. And that is going to be useful in our next experiment, where we look at the way capacitors are affected by a.c. The circuit we'll look at is shown in Figure 5.5 and is very simple, but it'll do nicely, thank you. It should remind you of a similar circuit we have already looked at, the voltage divider, only one of the two resistors of the voltage divider has been replaced by a capacitor. Like an ordinary voltage divider the circuit has an input and an output. What we're going to attempt to do in the experiment is to measure the output signal when the input signal is supplied from our astable multi-vibrator.

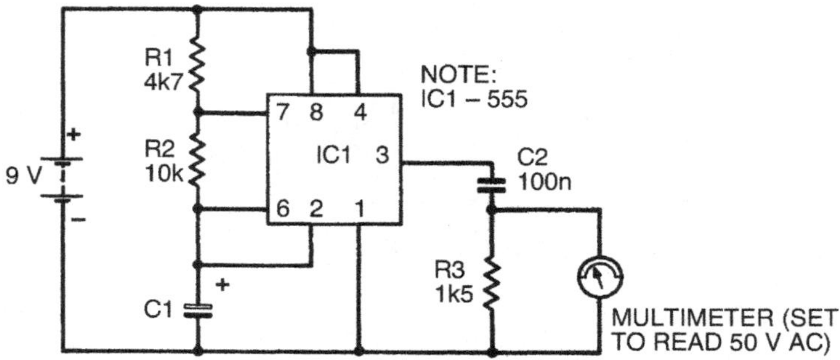

FIGURE 5.6 A circuit combining the astable multi-vibrator and the circuit in Figure 5.5.

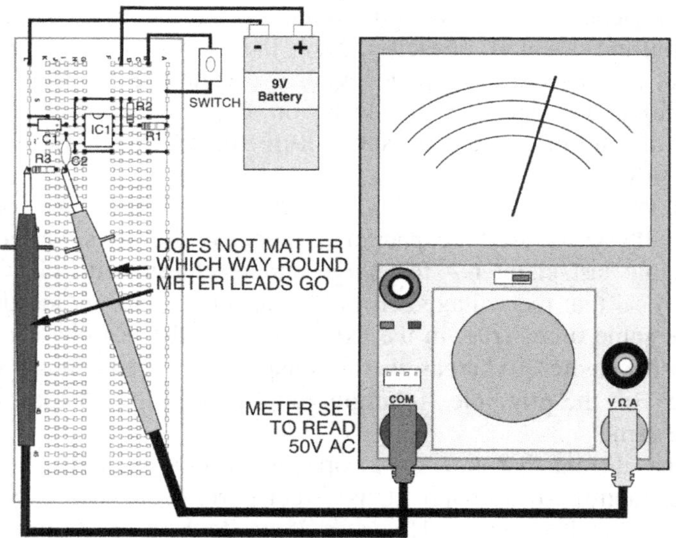

FIGURE 5.7 The breadboard layout of the circuit in Figure 5.6. It is the same as that in Figure 5.4, with a capacitor in place of the LED.

Figure 5.6 shows the whole circuit of the experiment while Figure 5.7 shows the breadboard layout. The procedure for the experiment is pretty straightforward: measure the output voltage of the a.c. voltage divider when a number of different frequencies are generated by the astable multi-vibrator, then tabulate and plot these results on a graph. Things really couldn't be easier. Table 5.1 is the table to fill in as you obtain your results and Figure 5.8 is marked out in a suitable grid to plot your graph. To change the

TABLE 5.1 Results When Capacitor C2 is 100 nF

Value of C1	Calculated Frequency	Measured Voltage
10 μF	5.8 Hz	
1 μF	58 Hz	
100 nF	580 Hz	
10 nF	5.8 kHz	
1 nF	58 kHz	

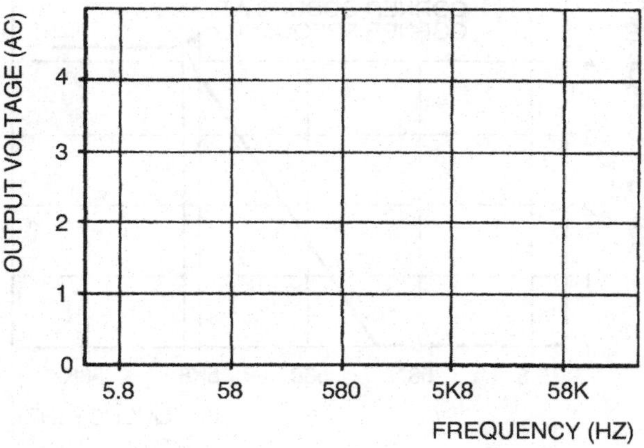

FIGURE 5.8 A blank graph on which you can plot your experimental results. Use Table 5.1 as well.

astable multi-vibrator's frequency, it is only necessary to change capacitor C1. Increasing it 10-fold decreases the frequency by a factor of 10; decreasing the capacitor value by 10 increases the frequency 10-fold. Five different values of capacitor therefore give an adequate range of frequencies.

As you do the experiment you'll find that only quite low voltages are measured (up to about 4 V a.c.) so, depending on your multimeter's lowest a.c. range, you may not achieve the level of accuracy you would normally desire, but the results will be OK, nevertheless.

Table 5.2 and Figure 5.9 show my results, which should be similar to yours (if not, you're wrong – I can't be wrong, can I?). The graph shows that the size of the output signal of the

TABLE 5.2 My Tabled Results

Value of C1	Calculated Frequency	Measured Voltage
10 μF	5.8 Hz	0 V
1 μF	58 Hz	0 V
100 nF	580 Hz	1.5 V
10 nF	5.8 kHz	4 V
1 nF	58 kHz	4.2 V

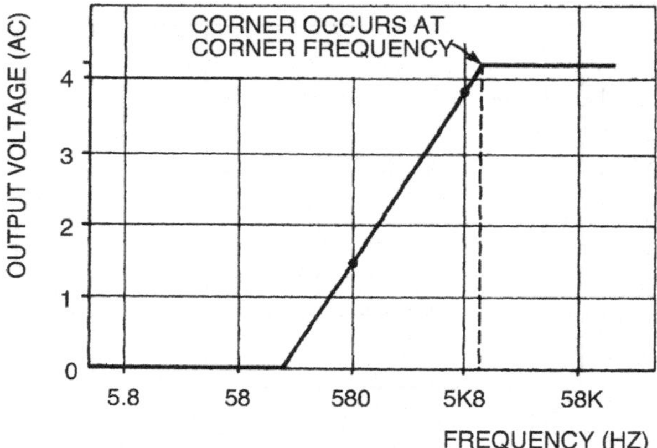

FIGURE 5.9 Graphed results of Table 5.2.

a.c. voltage divider is dependent on the frequency of the applied input signal. In particular, there are three clearly distinguishable sections to this graph, each relating to frequency:

- First, above a certain frequency, known as the corner frequency, the output signal is constant and at its maximum.
- Second, at low frequencies (close to 0Hz) the output is zero.
- Third, between these two sections the output signal varies in size depending on the applied input signal frequency.

Is this the same for all a.c. voltage dividers of the type shown in Figure 5.5? Well, let's repeat the experiment, using a different capacitor for C2, to find out.

TABLE 5.3 Results When Capacitor C2 is 10 nF

Value of C1	Calculated Frequency	Measured Voltage
10 μF	5.8 Hz	
1 μF	58 Hz	
100 nF	580 Hz	
10 nF	5.8 kHz	
1 nF	58 kHz	

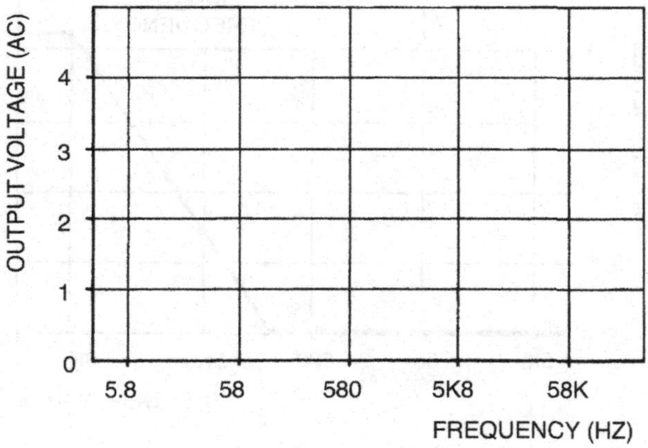

OUTPUT VOLTAGE (AC)
4
3
2
1
0
5.8 58 580 5K8 58K
FREQUENCY (HZ)

FIGURE 5.10 A graph for use with your results from the experiment with the 10 nF capacitor in Table 5.3.

Try a 10 nF capacitor and repeat the whole procedure, putting your results in Table 5.3 and Figure 5.10. Table 5.4 and Figure 5.11 show our results.

And yes, the graph is the same shape but is moved along the horizontal axis by an amount equivalent to a 10-fold increase in frequency (the capacitor was decreased in value by 10-fold, remember). A similar inverse relationship is caused by changing the resistor value too.

Frequency, capacitance, and resistance are related in the a.c. voltage divider by the expression:

$$f = \frac{1}{RC}$$

TABLE 5.4 My Results (C2 = 10 nF)

Value of C1	Calculated Frequency	Measured Voltage
10 μF	5.8 Hz	0 V
1 μF	58 Hz	0 V
100 nF	580 Hz	0 V
10 nF	5.8 kHz	1.5 V
1 nF	58 kHz	4 V

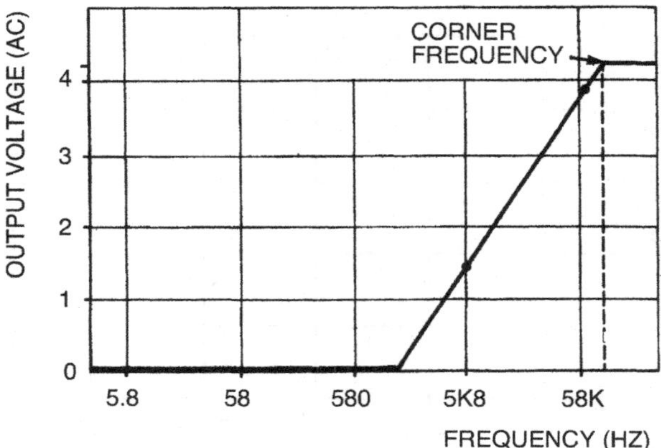

FIGURE 5.11 Graph of my results from Table 5.4 in the same experiment.

where f is the corner frequency. For the first voltage divider, with a capacitance of 100 nF and a resistance of 1.5 kΩ, the corner frequency is:

$$f = \frac{1}{1500 \times 100 \times 10^{-9}}$$

$$= 6666 \text{ Hz}$$

which is more or less what we found in the experiment. In the second voltage divider, with a 10 nF capacitor, the corner frequency increases by 10 to 66,666 Hz.

Remembering what we learned in the last chapter about resistors and capacitors in charging/discharging circuits, we can simplify the expression for corner frequency to:

$$f = \frac{1}{\tau}$$

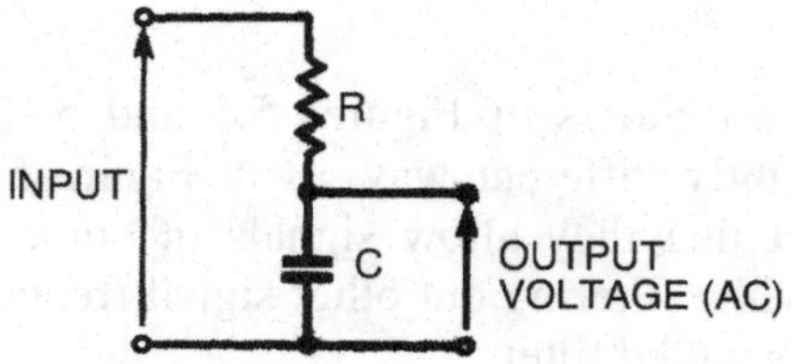

FIGURE 5.12 An a.c. voltage divider with the resistor and capacitor transposed.

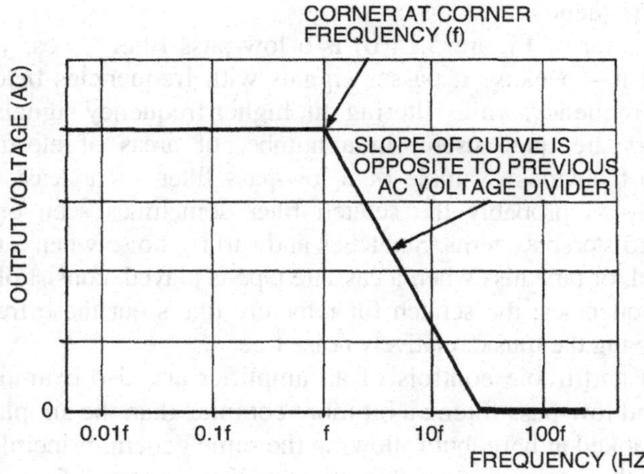

FIGURE 5.13 This graph shows the similarities and differences between this circuit and that in Figure 5.5.

because the product *RC* is the time constant, τ. This may be easier for you to remember.

An a.c. voltage divider can be constructed in a different way, as shown in Figure 5.12. Here the resistor and capacitor are transposed. What do you think the result will be? Well, the output signal size now decreases with increasing frequency – exactly the opposite effect of the a.c. voltage divider of Figure 5.5! All other aspects are the same, however: there is a constant section below a corner frequency, and a section where the output signal is zero, as shown in Figure 5.13. Once again the corner frequency is given by the expression:

$$f = \frac{1}{RC} = \frac{1}{\tau}$$

FILTER TIPS

The a.c. voltage dividers of Figures 5.5 and 5.12 are normally shown in a slightly different way, as in Figure 5.14(a) and (b). Due to the fact that they allow signals of some frequencies to pass through, while filtering out other signal frequencies, they are more commonly called filters.

The filter of Figure 5.14(a) is known as a high-pass filter – because it allows signal frequencies higher than its corner frequency to pass while filtering out signal frequencies lower than its corner frequency.

The filter of Figure 5.14(b) is a low-pass filter – yes, you've guessed it – because it passes signals with frequencies below its corner frequency, while filtering out higher frequency signals.

Filters are quite useful in a number of areas of electronics. The most obvious example of a low-pass filter — at least for us oldies — is probably the scratch filter sometimes seen on old-fashioned stereo systems. Scratches and surface noise when a record is played, or tape hiss when a cassette tape is played, consist of quite high frequencies; the scratch filter merely filters out these frequencies, leaving the music relatively noise free.

Bass and treble controls of an amplifier are also examples of high- and low-pass filters: a bit more complex than the simple ones we've looked at here, but following the same general principles.

And that's about it for this chapter. You can try a few experiments of your own with filters if you want. Just remember that whenever you use your meter to measure voltage across a resistor in a filter, the meter resistance affects the actual value of resistance and can thus drastically affect the reading.

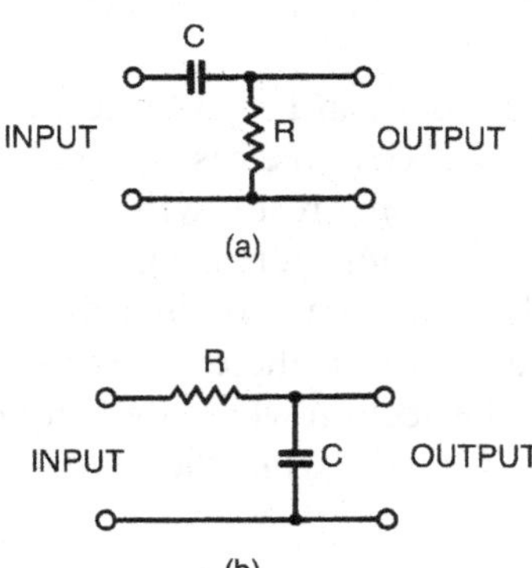

FIGURE 5.14 The a.c. divider sections of the circuits in Figures 5.5 and 5.12.

Now try the quiz that follows to check and see if you've been able to understand what we've been looking at in this chapter.

QUIZ

Answers at the end of the book.

1. A signal of frequency 1 kHz is applied to a low-pass filter with a corner frequency of 10 kHz. What happens?:
   a. The output signal is one-tenth of the input signal
   b. The output signal is larger than the input signal
   c. There is no output signal
   d. The output signal is identical to the input signal
   e. All of these.
2. A high-pass filter consisting of a 10 kΩ resistor and an unknown capacitor has a corner frequency of 100 Hz. What is the value of the capacitor?:
   a. 1 nF
   b. 10 nF
   c. 100 nF
   d. 1000 nF
   e. 1 μF.
3. A low-pass filter consisting of a 1 μF capacitor and an unknown resistor has a corner frequency of 100 Hz. What is the value of the resistor?:
   a. 10 kΩ
   b. 1 kΩ
   c. 100 kΩ
   d. All of these
   e. It makes no difference
   f. d and e
   g. None of these.
4. In a circuit similar to that in Figure 5.2, resistor R1 is 10 kΩ, resistor R2 is 100 kΩ, and capacitor C1 is 10 nF. The output frequency of the astable multi-vibrator is:
   a. A square wave
   b. About 680 Hz
   c. Too fast to see the LED flashing
   d. a and b
   e. c and d
   f. None of these.

Chapter 6

Diodes I

We're going to take a close look at a new type of component in this chapter – the diode. Diodes are the simplest component in the range of devices known as semiconductors. Actually, we've briefly looked at a form of diode before – the light-emitting diode, or LED – and we've seen another semiconductor too: the 555 integrated circuit. But now we'll start to consider semiconductors in depth.

Naturally, you'll need some new components for the circuits you're going to build here. These are:

- 1 × 150 Ω, 0.5 W resistor
- 1 × 1N4001 diode
- 1 × OA47 diode
- 1 × 3V0 zener diode (type BZY88)
- 1 × 1k0 miniature horizontal preset.

Diodes get their name from the basic fact that they have two electrodes (di-ode, geddit?). One of these electrodes is known as the anode, the other is the cathode. Figure 6.1 shows the symbol for a diode, where the anode and cathode are marked. Figure 6.2 shows some typical diode body shapes, again with the cathode marked.

Photo 6.1 is a photograph of a miniature horizontal preset resistor. We're going to use it in the following circuits as a variable voltage divider. To adjust it you'll need a small screwdriver or tool to fit in the adjusting slot – turning it one way and another alters the position of the preset's wiper over the resistance track.

Figure 6.3 shows the circuit we're going to build first in this chapter. It's very simple, using two components we're already familiar with (a resistor and an LED), together with the new component we want to look at: a diode. Before you build it, note which way round the diode is and also make sure you get the LED polarized correctly too. In effect, the anodes of each diode

FIGURE 6.1 The circuit symbol for an ordinary diode.

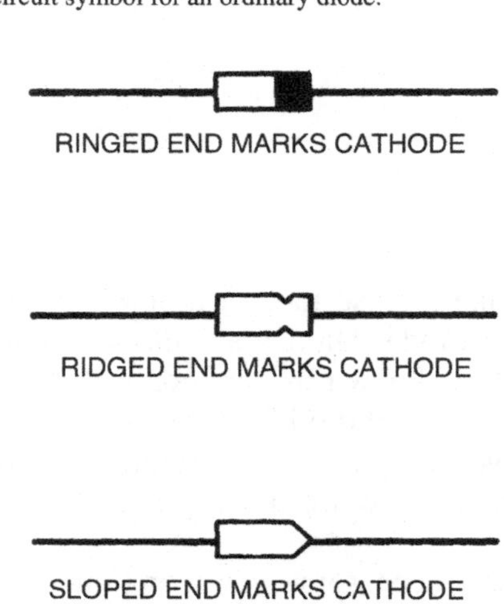

FIGURE 6.2 Some typical diode body shapes.

PHOTO 6.1 A horizontal preset resistor.

(an LED is a diode, too, remember – a light-emitting diode) connect to the more positive side in the circuit. A breadboard layout is shown in Figure 6.4, though by this stage you should perhaps be confidently planning your own breadboard layouts.

WHICH WAY ROUND?

If you've connected the circuit up correctly, the LED should now be on. This proves that current is flowing. To calculate exactly

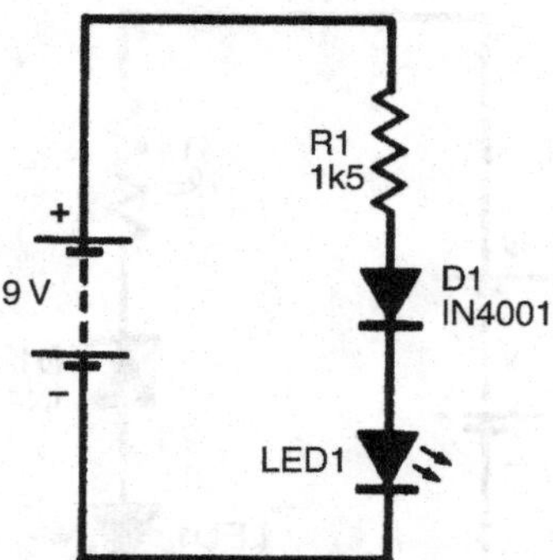

FIGURE 6.3 Our first simple circuit using a diode.

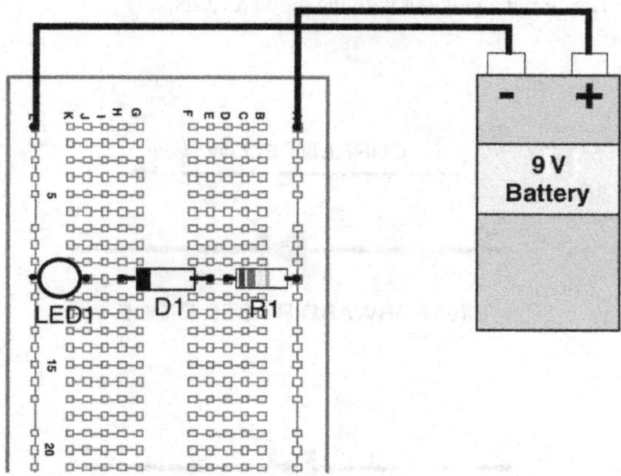

FIGURE 6.4 A breadboard layout for the circuit in Figure 6.3.

what current, we can use Ohm's law. Let's assume that the total battery voltage of 9V is dropped across the resistor and that no voltage occurs across the two diodes. In fact, there is voltage across the diodes, but we needn't worry about it yet, as it is only a small amount. However, we'll measure it soon.

Now, with a resistance of 1k5 and a voltage of 9V, we can calculate the current flowing as:

$$I = \frac{V}{R} = \frac{9}{1500} = 6\,\text{mA}$$

The next thing to do is to turn around the diode, so that its cathode is more positive, as in the circuit of Figure 6.5. The

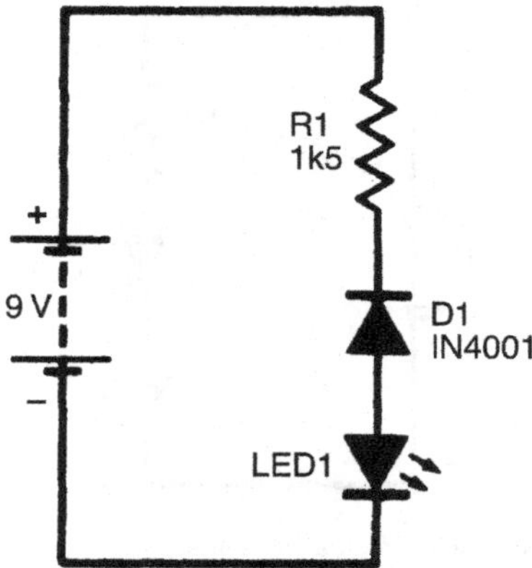

FIGURE 6.5 The circuit again, but with the diode reversed.

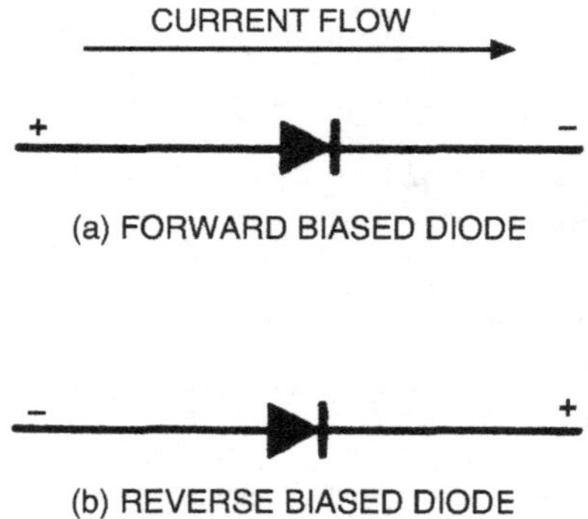

FIGURE 6.6 Circuit diagrams for forward- and reverse-biased diodes.

breadboard layout is the same with anode and cathode reversed, so we needn't redraw it.

What happens? You should find that absolutely nothing happens. The LED does not light up, so no current must be flowing. The action of reversing the diode has resulted in the stopping of current. We can summarize this quite simply in Figure 6.6.

Figure 6.6(a) shows a diode whose anode is positive with respect to its cathode. Although we've shown the anode as positive with a + symbol, and the cathode as negative with a − symbol, they don't necessarily have to be positive and negative. The cathode could, for example, be at a voltage of +1000 V if the anode was at a greater positive voltage of, say, +1001 V. All that needs to occur is that the anode is positive with respect to the cathode.

Under such a condition, the diode is said to be forward biased and current will flow from anode to cathode.

When a diode is reverse biased, i.e. its cathode is positive with respect to the anode, no current flows, as shown in Figure 6.6(b). Obviously, something happens within the diode that we cannot see, depending on the polarity of the applied voltage to define whether current can flow or not. It isn't necessary to understand here just exactly what this something is. We needn't know any more about it here because we're only concerned with the practical aspects at the moment, and all we need to remember is that a forward-biased diode conducts, allowing current to flow, while a reverse-biased diode doesn't.

What we do need to consider in more detail, however, is the value of the current flowing, and the small, but nevertheless apparent, voltage that occurs across the diode, when a diode is forward biased (the voltage we said earlier we needn't then worry about). The following experiment will show how the current and the voltage are related.

Figure 6.7 shows the circuit you have to build. You'll see that two basic measurements need to be taken with your meter. The first measurement is the voltage across the forward-biased diode, the second measurement is the current through it. Each measurement needs to be taken a number of times as the preset is varied in an organized way. Table 6.1, which is half complete, is for you to

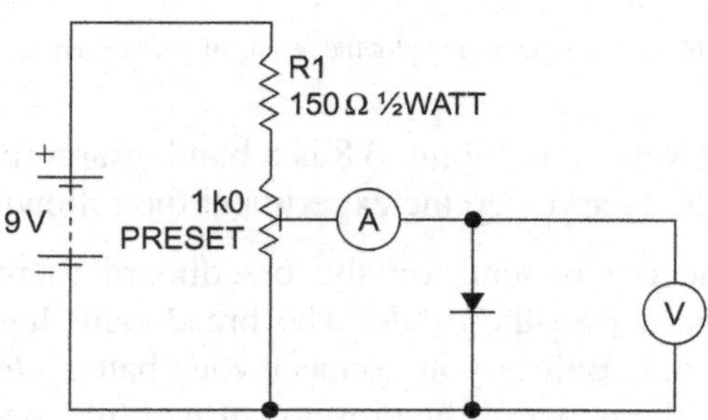

FIGURE 6.7 A circuit to test the operation of a forward-biased diode.

TABLE 6.1 This is Half Complete; Add the Results of Your Experiment

Current (mA)	Voltage (V)
0	0
	0.4
	0.6
2	
5	
10	
20	

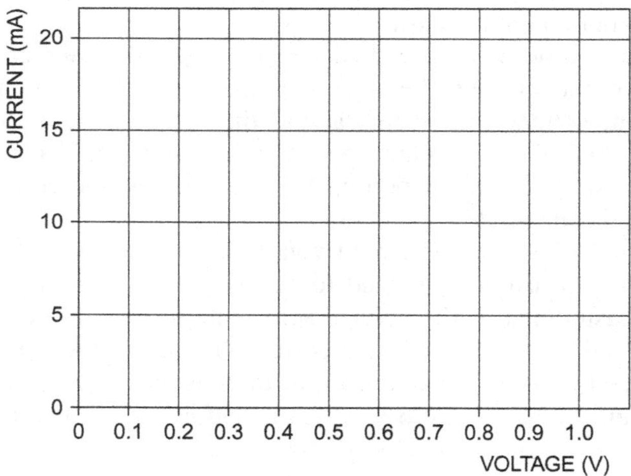

FIGURE 6.8 A blank graph for you to plot the results of your experiment.

record your results, and Figure 6.8 is a blank graph for you to plot the results into a curve. Do the experiment the following way:

1. Set up the components on the breadboard to measure only the voltage across the diode. The breadboard layout is given in Figure 6.9. Before you connect your battery to the circuit, make sure the wiper of the preset is turned fully anticlockwise.
2. Adjust the preset wiper clockwise, until the first voltage in Table 6.1 is reached.

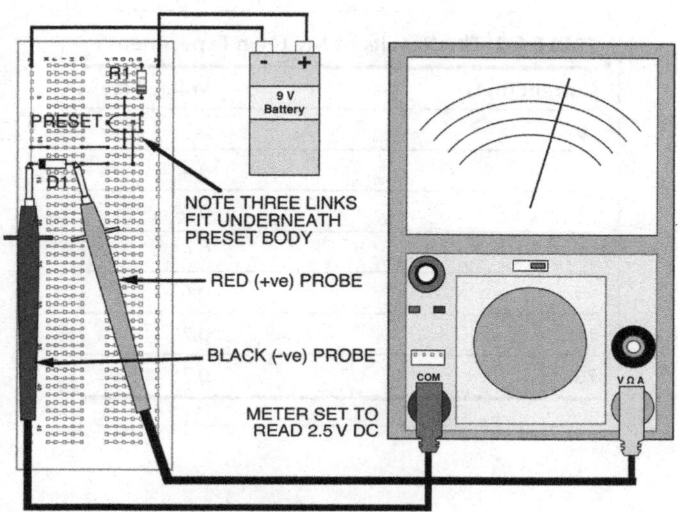

FIGURE 6.9 The breadboard layout for the circuit in Figure 6.7.

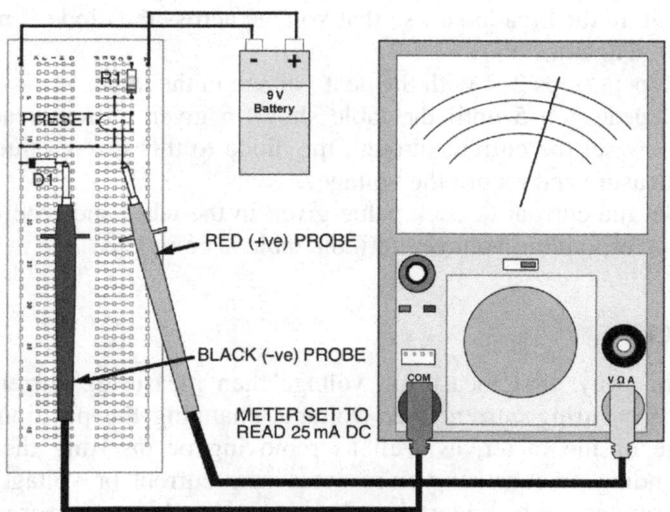

FIGURE 6.10 The same circuit, set up to measure the current through the diode.

3. Now set up the breadboard layout of Figure 6.10, to measure the current through the diode – the breadboard layout is designed so that all you have to do is take out a short length of single-strand connecting wire and change the position of the

TABLE 6.2 The Results of My Own Experiment

Current (mA)	Voltage (V)
0	0
0	0.4
1	0.6
2	0.65
5	0.65
10	0.7
20	0.75

meter and its range. Record the value of the current at the voltage of step 2.

4. Change the position of the meter and its range, and replace the link in the breadboard so that voltage across the diode is measured again.
5. Repeat steps 2–4 with the next voltage in the table.
6. Repeat step 5 until the table shows a given current reading. Now set the current through the diode to this given value and measure and record the voltage.
7. Set the current to each value given in the table and record the corresponding voltage, until the table is complete.

TRICKY

In this way, first measuring voltage then measuring current, or first measuring current then voltage, changing the position and range of the meter, as well as removing or inserting the link depending on whether you're measuring current or voltage, the experiment can be undertaken. Yes, it's tricky, but we never said it was a doddle, did we? You'll soon get the hang of it and get some good results.

Now plot your results on the graph of Figure 6.8. My results (correct naturally!) are shown in Table 6.2 and Figure 6.11.

Repeat the whole experiment again, using the OA47 diode this time. You can put your results in Table 6.3 and plot your graph in Figure 6.12. Our results are in Table 6.4 and Figure 6.13.

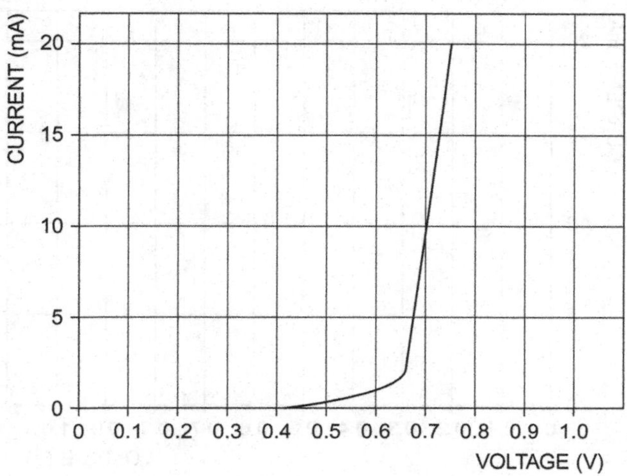

FIGURE 6.11 My own results from the experiment.

TABLE 6.3 The Results of Your Experiment

Current (mA)	Voltage (V)
0	0
	0.1
	0.2
	0.25
	0.3
3	
5	
10	
20	

As you might expect, these two plotted curves are the same basic shape. The only real difference between them is that they change from a level to an extremely steep line at different positions. The OA47 curve, for example, changes at about 0V3, while the 1N4001 curve changes at about 0V65.

The sharp changes in the curves correspond to what are sometimes called transition voltages – the transition voltage for the OA47 is about 0V3; the transition voltage for the 1N4001 is about 0V65. It's important to remember, though, that the transition

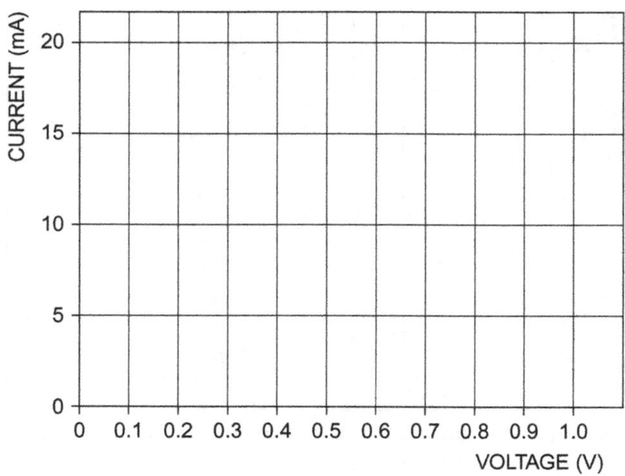

FIGURE 6.12 Use this graph to plot your results from the second experiment.

TABLE 6.4 My Results From the Second Experiment

Current (mA)	Voltage (V)
0	0
0	0.1
0.5	0.2
1	0.25
2	0.3
3	0.32
5	0.35
10	0.4
20	0.45

voltages in these curves are only for the particular current range under consideration – 0 to 20 mA in this case. If similar curves are plotted for different current ranges then slightly different transition voltages will be obtained. In any current range, however, the transition voltages won't be more than about 0V1 different to the transition voltages we've seen here. The two curves – of

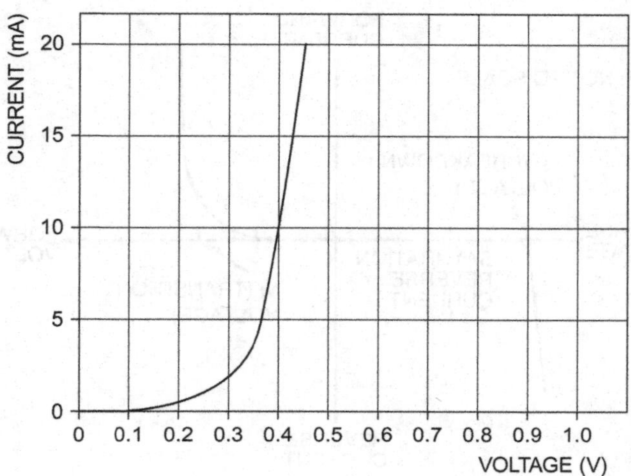

FIGURE 6.13 A graph plotting my results from the second experiment.

the OA47 and the 1N4001 diodes – show that a different transition voltage is obtained (0V3 for the OA47, 0V65 for the 1N4001) depending on which semiconductor material a diode is made from. The OA47 diode is made from germanium while the 1N4001 is of silicon construction. All germanium diodes have a transition voltage of about 0V2 to 0V3; similarly all silicon diodes have a transition voltage of about 0V6 to 0V7.

These two curves are exponential curves – in the same way that capacitor charge/discharge curves (see Chapter 4) are exponential, but in a different direction, that's all – and form part of what are called diode characteristic curves or sometimes simply diode characteristics. But the characteristics we have determined here are really only half the story as far as diodes are concerned. All we have plotted are the forward voltages and resultant forward currents when the diodes are forward biased. If diodes were perfect this would be all the information we need. But – yes you've guessed it, diodes are not perfect – when they are reverse biased so that they have reverse voltages, reverse currents flow. So, to get a true picture of diode operation we have to extend the characteristic curves to include reverse-biased conditions.

Reverse biasing a diode means that its anode is more negative with respect to its cathode. So by interpolating the *x*- and *y*-axes of the graph, we can provide a grid from the diode characteristic that allows it to be drawn in both forward- and reverse-biased conditions.

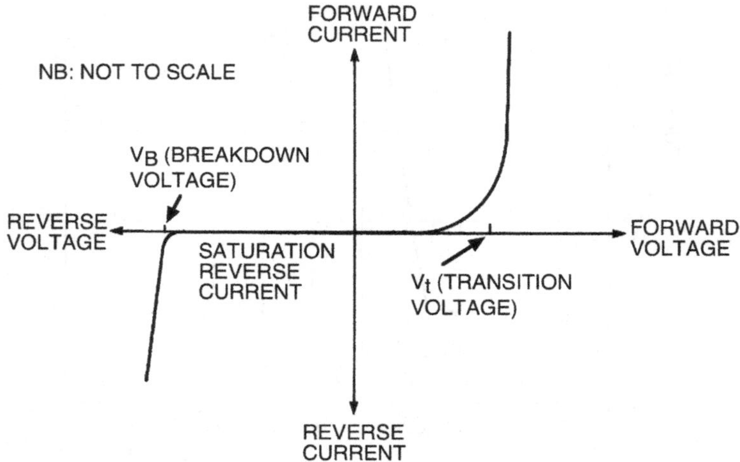

FIGURE 6.14 Plotting the reverse-bias characteristics for an ordinary diode is not practical, so we give you the characteristic curve here.

It wouldn't be possible for you to plot the reverse-biased conditions, for an ordinary diode, the way you did the forward-biased experiment (we will, however, do it for a special type of diode soon), so instead we'll make it easy for you and give you the whole characteristic curve. Whatever type of diode, it will follow a similar curve to that of Figure 6.14, where the important points are marked.

REVERSE BIAS

From the curve you can see that there are two distinct parts which occur when a diode is reverse biased. First, at quite low reverse voltages, from about −0.V1 to the breakdown voltage, there is a more or less constant but small reverse current. The actual value of this reverse current (known as the saturation reverse current, or just the saturation current) depends on the individual diode, but is generally of the order of microamps.

The second distinct part of the reverse-biased characteristic occurs when the reverse voltage is above the breakdown voltage. The reverse current increases sharply with only comparatively small increases in reverse voltage. The reason for this is because of electronics breakdown of the diode when electrons gain so much energy due to the voltage that they push into one another, just like rocks and boulders rolling down a steep mountainside

FIGURE 6.15 The circuit symbol for a zener diode.

Hint

The way to remember the zener diode circuit symbol is to note that the bent line representing its cathode corresponds to the electronics breakdown.

push into other rocks and boulders, which in turn start to roll down the mountainside, pushing into more rocks and boulders, forming an avalanche. This analogy turns out to be an apt one, and in fact the electronic diode breakdown voltage is sometimes referred to as avalanche breakdown and the breakdown voltage is sometimes called the avalanche voltage. Similarly the sharp knee in the curve at the breakdown voltage is often called the avalanche point.

In most ordinary diodes the breakdown voltage is quite high (in the 1N4001 it is well over −50V), so this is one reason why you couldn't plot the whole characteristic curve, including reverse-biased conditions, in the same experiment – our battery voltage of 9V simply isn't high enough to cause breakdown.

Some special diodes, on the other hand, are purposefully manufactured to have a low breakdown voltage, and you can use one of them to study and plot your own complete diode characteristic. Such diodes are named after the American scientist, Zener, who was one of the first people to study electronic breakdown. The zener diode you are going to use is rated at 3V0, which of course means that its breakdown occurs at 3V, which is below the battery voltage and is therefore plottable in the same sort of experiment as the last one. The symbol for a zener diode, incidentally, is shown in Figure 6.15.

The procedure is more or less the same as before. The circuit is shown in Figure 6.16, with the zener diode shown as forward biased. Complete Table 6.5 with the circuit as shown, then turn the zener diode round as shown in the circuit of Figure 6.17 (this is the way zener diodes are normally used) so that it is reverse biased, then perform the experiment again, completing Table 6.6 as you go along. Although all the results will in theory be

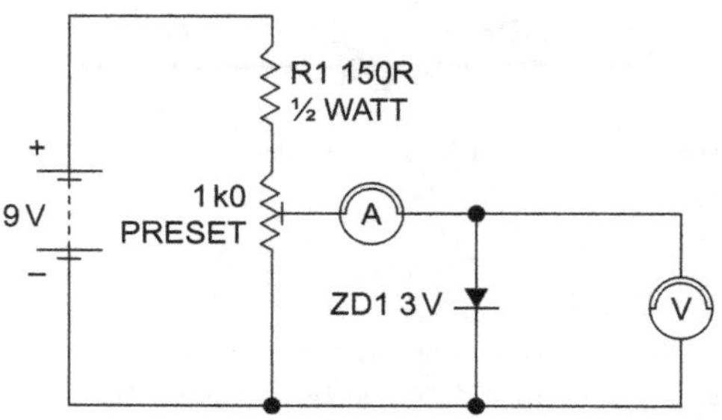

FIGURE 6.16 A circuit with the zener diode forward biased.

TABLE 6.5 The Results of Your Experiment

Current (mA)	Voltage (V)
0	0
	0.4
	0.6
1	
2	
5	
10	
20	

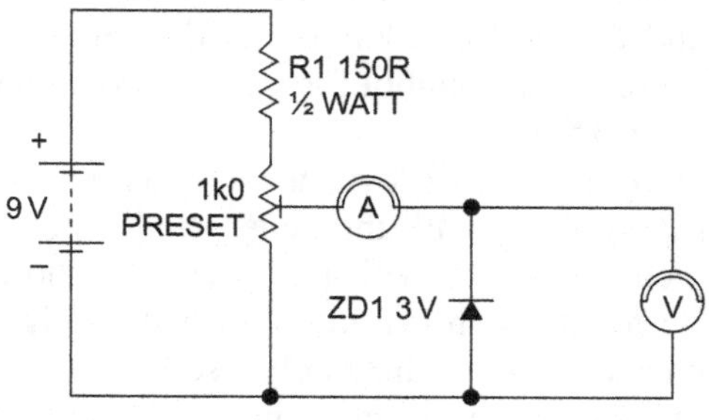

FIGURE 6.17 The circuit with the zener diode reverse biased.

TABLE 6.6 The Results of Your Further Experiments

−Current (mA)	−Voltage (V)
0	0
	1
	2
	2.5
	3
10	
20	

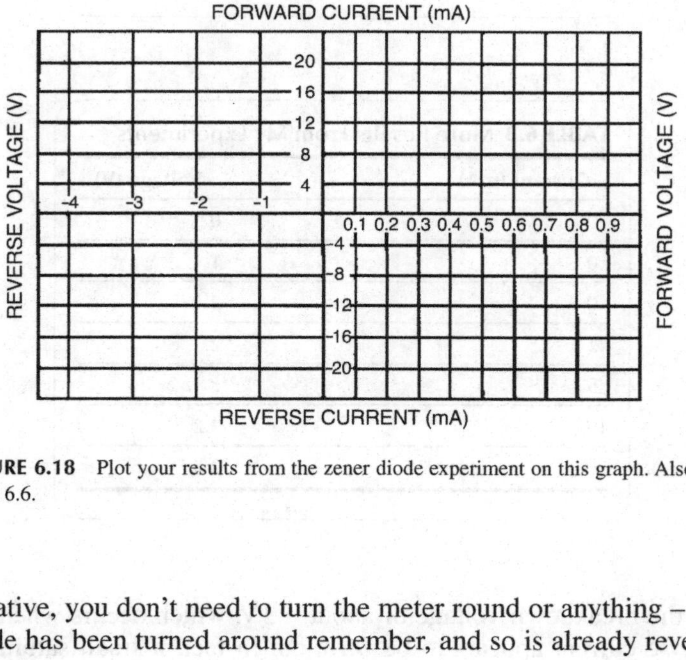

FIGURE 6.18 Plot your results from the zener diode experiment on this graph. Also use Table 6.6.

negative, you don't need to turn the meter round or anything – the diode has been turned around remember, and so is already reverse biased.

Next, plot your complete characteristic on the graph of Figure 6.18. My results and characteristics are given in Tables 6.7 and 6.8, and Figure 6.19. Yours should be similar.

From the zener diode characteristic you will see that it acts like any ordinary diode. When forward biased it has an exponential curve with a transition voltage of about 0V7 for the current range observed. When reverse biased, on the other hand, you can

TABLE 6.7 The Results of My Experiments

Current (mA)	Voltage (V)
0	0
0	0.4
0	0.6
1	0.7
2	0.75
5	0.75
10	0.8
20	0.8

TABLE 6.8 More Results From My Experiments

−Current (mA)	−Voltage (V)
0	0
0	1
0.5	2
2	2.5
5	3
10	3.2
20	3.5

see the breakdown voltage of about −3 V, which occurs when the reverse current appears to be zero, but in fact, a small saturation current does exist – it was simply too small to measure on the meter.

Voilà! – A Complete Diode Characteristic Curve

Finally, it stands to reason that any diode must have maximum ratings above which the heat generated by the voltage and current is too much for the diode to withstand. Under such circumstances

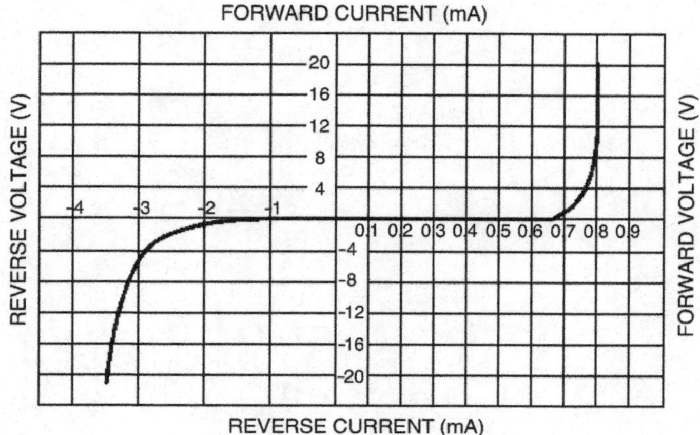

FIGURE 6.19 My results from the zener diode experiment are given in this graph and in Tables 6.7 and 6.8.

TABLE 6.9 Typical Maximum Ratings of the Two Diodes We Have Been Looking At

Maximum Ratings	1N4001	OA47
Maximum mean forward current	1 A	110 mA
Maximum repetitive forward current	10 A	150 mA
Maximum reverse voltage	−50 V	−25 V
Maximum operating temperature	170°C	60°C

the diode body may melt (if it is a glass diode such as the OA47) or, more likely, it will crack and fall apart. To make sure their diodes don't encounter such rough treatment manufacturers supply maximum ratings that should not be exceeded. Typical maximum ratings of the two ordinary diodes we have looked at, the 1N4001 and the OA47, are listed in Table 6.9.

In the next chapter we'll be considering diodes again, and how we can use them, practically, in circuits, what their main uses are, and how to choose the best one for any specific purpose.

Chapter 7

Diodes II

In the last chapter, we took a detailed look at diodes and their characteristic curves. In this chapter we're going to take this one stage further and consider how we use the characteristic curve to define how the diode will operate in any particular circuit. Finally, we'll look at a number of circuits that show some of the many uses of ordinary diodes.

The components you'll need for the circuits in this chapter are:

- 1 × 10k resistor
- 2 × 15k resistors
- 2 × 100k resistors
- 1 × 10μF 10V electrolytic capacitor
- 1 × 220μF 10V electrolytic capacitor
- 1 × 1N4001 diode
- 1 × LED
- 1 × 555 integrated circuit.

Figure 7.1 shows the forward-biased section of a typical diode characteristic curve. It has a transition voltage of about 0V7, so you should know that it's the characteristic curve of a silicon diode.

Diodes aren't the only electronic components for which characteristic curves may be drawn – most components can be studied in this way. After all, the curve is merely a graph of the voltage across the component compared to the current through it. So, it's equally possible that we draw a characteristic curve of, say, a resistor. To do this we could perform the same experiment we did in the last chapter with the diodes: measuring the voltage and current at a number of points, then sketching the curve as being the line that connects the points marked on the graph.

But there's no need to do this in the case of a resistor, because we know that resistors follow Ohm's law. We know that:

$$R = \frac{V}{I}$$

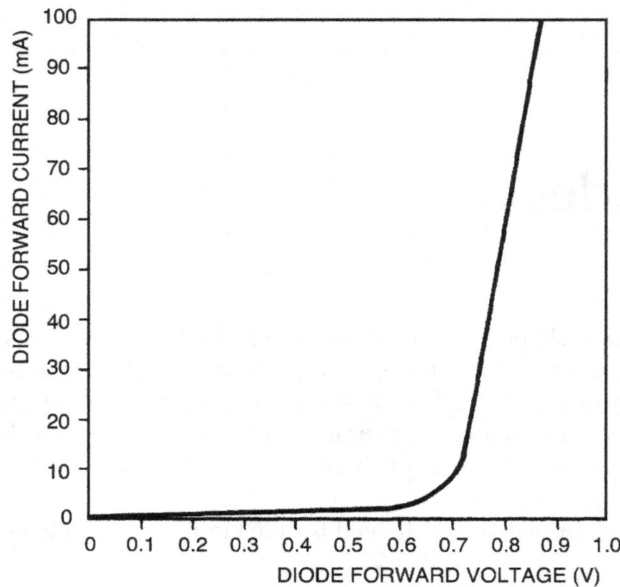

FIGURE 7.1 The forward-biased section of the characteristic curve of a silicon diode.

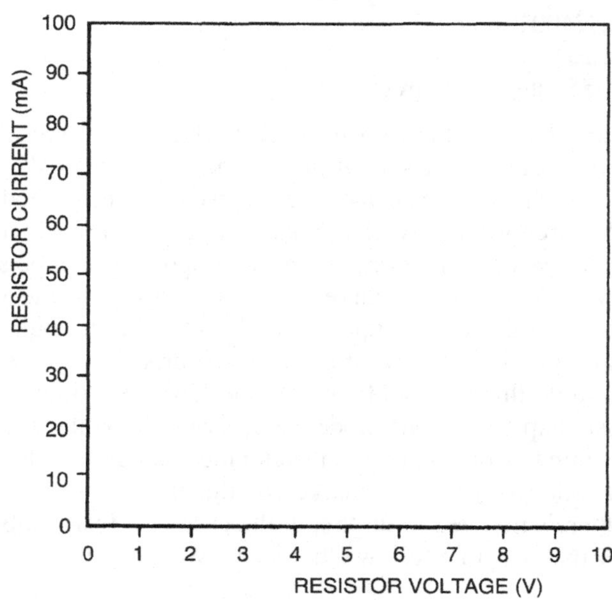

FIGURE 7.2 A blank graph for you to fill in – see the text above for instructions.

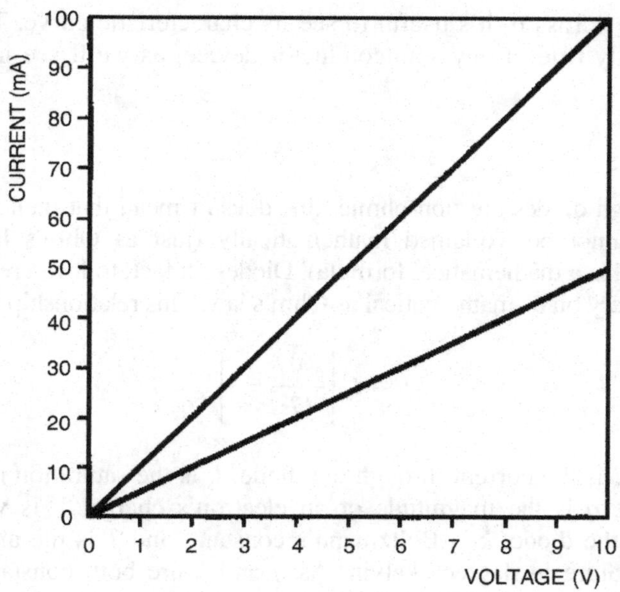

FIGURE 7.3 Your efforts with Figure 7.2 should produce a graph something like this.

Take Note

Resistor characteristic curves are straight lines because resistors follow Ohm's law (we say they are ohmic) and Ohm's law is a linear relationship. So resistor characteristic curves are linear too. And because they are linear there's really no point in drawing them, and you'll never see them in this form anywhere else – we drew them simply to emphasize the principle.

where R is the resistance, V is the voltage across the resistor, and I is the current through it. So, for any value of resistor, we can choose a value for, say, the voltage across it, and hence calculate the current through it. Figure 7.2 is a blank graph. Calculate and then draw on the graph characteristic curves for two resistors of values 100 and 200 Ω. The procedure is simple, just calculate the current at each voltage point for each resistor.

Your resultant characteristic curves should look like those in Figure 7.3 – two straight lines.

We need to draw diode characteristic curves, on the other hand, because they're non-ohmic and hence non-linear. So to see what current passes through the device with any particular voltage across it, it's useful to see its characteristic curve. This is generally true of any semiconductor device, as we'll see in later chapters.

MATHS

Although diodes are non-ohmic, this doesn't mean that their operation cannot be explained mathematically (just as Ohm's law, or $V = IR$, is a mathematical formula). Diodes, in fact, follow a relationship every bit as mathematical as Ohm's law. This relationship is:

$$I = I_s \left\{ \frac{e^{(qV)}}{kT} - 1 \right\}$$

where I is the current through the diode, I_s is the saturation reverse current, q is the magnitude of an electron's charge, V is voltage across the diode, k is Boltzmann's constant, and T is the absolute temperature in degrees kelvin. As q and k are both constant and at room temperature the absolute temperature is more or less constant, the part of the equation q/kT is also more or less constant at about 40 (work it out yourself if you want: q is 1.6×10^{-19} C, k is 1.38×10^{-23} J K^{-1}, and room temperature, say 17°C, is 290 K).

The equation is thus simplified to be approximately:

$$I = I_s(e^{40V} - 1)$$

The exponential factor (e^{40V}), of course, confirms what we already knew to be true – that the diode characteristic curve is an exponential curve. From this characteristic equation we may calculate the current flowing through a diode for any given voltage across it, just as the formulae associated with Ohm's law do for resistors.

Hint

But even with this simplified approximation of the characteristic equation, you can appreciate the value of having a characteristic curve in front of you to look at. If I had the option between having to use the equation or a diode characteristic curve to calculate the current through a diode, I know which I'd choose!

LOAD LINES

It's important to remember that although a diode characteristic curve is non-linear and non-ohmic, so that it doesn't abide by Ohm's law throughout its entire length, it does follow Ohm's law at any particular point on the curve. So, say, if the voltage across the diode whose characteristic curve is shown in Figure 7.1 is 0V8, so the current through it is about 80 mA (check it yourself), then the diode resistance is:

$$R = \frac{V}{I} = \frac{0.8}{80 \times 10^{-3}} = 10\,\Omega$$

as defined by Ohm's law. Any change in voltage and current, however, results in a different resistance.

Generally, diodes don't exist in a circuit merely by themselves. Other components, e.g. resistors, capacitors, and other components in the semiconductor family, are combined with them. It is when designing such circuits and calculating the operating voltages and currents in the circuits that the use of diode characteristic curves really comes in handy. Figure 7.4 shows as an example a simple circuit consisting of a diode, a resistor, and a battery. By looking at the circuit we can see that a current will flow. But what is this current? If we knew the voltage across the resistor we could calculate (from Ohm's law) the current through it, which is of course the circuit current. Similarly, if we knew the voltage across the diode we could determine (from the characteristic curve) the circuit current. Unfortunately we know neither voltage!

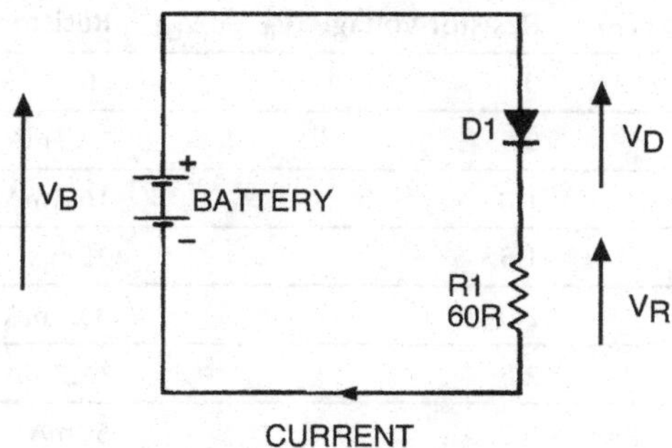

FIGURE 7.4 A simple diode circuit – but what is the current flow?

We do know, however, that the voltages across both the components must add up to the battery voltage. In other words:

$$V_B = V_D + V_R$$

(it's a straightforward voltage divider). This means that we can calculate each voltage as being a function of the battery voltage, given by:

$$V_D = V_B - V_R$$

and

$$V_R = V_B - V_D$$

Now, we know that the voltage across the diode can only vary between about 0V and 0V8 (given by the characteristic curve), but there's nothing to stop us hypothesizing about larger voltages than this, and drawing up a table of voltages that would thus occur across the resistor. Table 7.1 is such a table, but it takes the process one stage further by calculating the hypothetical current through the resistor at these hypothetical voltages.

From Table 7.1 we can now plot a second curve on the diode characteristic curve, of diode voltage against resistor current. Figure 7.5 shows the completed characteristic curve (labeled Load line R = 60R). The curve is actually a straight line – fairly obvious, if you think about it, because all we've done is plot a

TABLE 7.1 Diode Characteristics

Diode Voltage (V_D)	Resistor Voltage ($V_B - V_D$)	Resistor Current (I)
3	0	0
2.5	0.5	8.3 mA
2.0	1.0	16.7 mA
1.5	1.5	25 mA
1.0	2.0	33.3 mA
0.5	2.5	41.7 mA
0	3.0	50 mA

voltage and a current for a resistor, and resistors are ohmic and linear. Because in such a circuit as that of Figure 7.4 the resistor is known as a load, i.e. it absorbs electrical power, the line on the characteristic curve representing diode voltage and resistor current is called the load line.

Where the load line and the characteristic curve cross is the operating point. As its name implies, this is the point representing the current through and voltage across the components in the circuit. In this example the diode voltage is thus 1 V and the diode current is 33 mA at the operating point.

If you think carefully about what we've just seen, you should see that the load line is a sort of resistor characteristic curve. It doesn't look quite like those of Figure 7.3, however, because the load line does not correspond to resistor voltage and resistor current, but diode voltage and resistor current – so it's a sort of inverse resistor characteristic curve (shown in Figure 7.5).

The slope of the load line (and thus the exact position of the circuit operating point) depends on the value of the resistor. Let's change the value of the resistor in Figure 7.4 to, say, 200 Ω. What is the new operating point? Draw the new load line corresponding to a resistor of value 200 Ω in Figure 7.5 to find out. You don't need to draw up a new table as in Table 7.1 – we know it's a straight line so we can draw it if we have only two points on the line. These two points can be when the diode voltage is 0 V (thus the resistor voltage, $V_R = V_B - V_D$, equals the battery voltage), and when the diode voltage equals the battery voltage the resistor current is zero. Figure 7.6 shows how your results should appear. The new operating point corresponds to a diode voltage of 0V8 and current of about 11 mA. We'll be looking at other examples of the uses of load lines when we look at other semiconductor devices in later chapters.

DIODE CIRCUITS

We're now going to look at some ways that diodes may be used practically in circuits. We already know that diodes may be used in circuits for practical purposes. We already know that diodes allow current flow in only one direction (ignoring saturation reverse current and zener current for the time being) and this is one of their main uses – to rectify alternating current (a.c.) voltages into direct current (d.c.) voltages. The most typical source of a.c. voltage we can think of is the 230 V mains supply to every

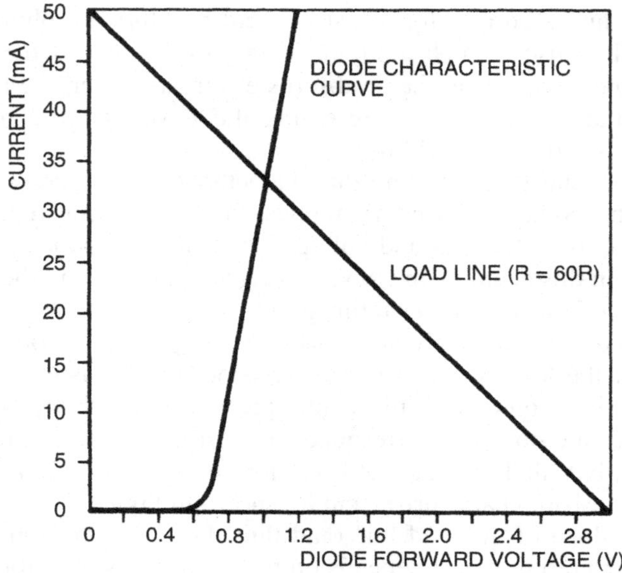

FIGURE 7.5 Load line from the results of Table 7.1.

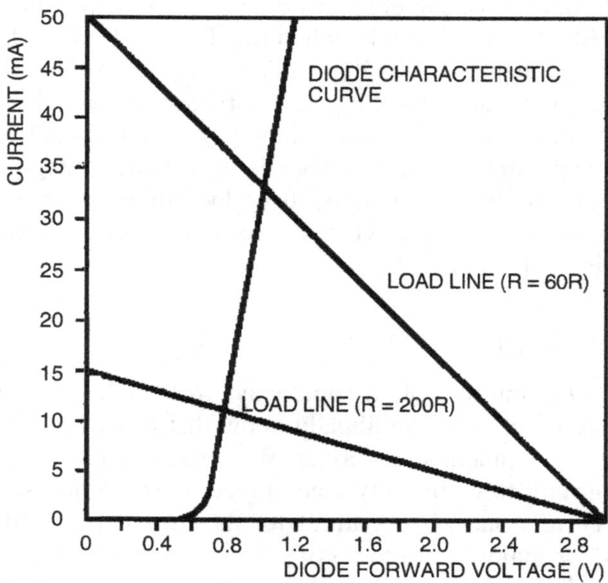

FIGURE 7.6 The new load line for a resistor of 200 Ω.

home in the UK. Most electronic circuits require d.c. power so we can understand that rectification is one of the most important uses of diodes. The part of any electronic equipment – TVs, radios, hi-fis, computers – that rectifies a.c. mains voltages into low d.c. voltages is known as the power supply (sometimes abbreviated to PSU – for power supply unit).

Generally, we wouldn't want to tamper with voltages as high as mains, so we would use a transformer to reduce the 230 V a.c. mains supply voltage to about 12 V a.c. We do not look at transformers in detail, it is suffice to know now that a transformer consists essentially of two coils of wire that are not in electrical contact. The circuit symbol of a transformer (Figure 7.7) shows this.

The simplest way of rectifying the a.c. output of a transformer is shown in Figure 7.8. Here a diode simply allows current to flow in one direction (to the load resistor, R_L), but not in the other direction (from the load resistor). The a.c. voltage from the transformer and the resultant voltage across the load resistor are shown in Figure 7.9. The resistor voltage, although in only one direction, is hardly the fixed voltage we would like, but nevertheless is technically a d.c. voltage. You'll see that, of

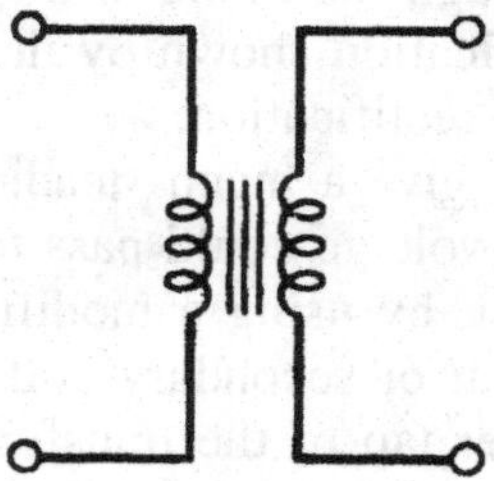

FIGURE 7.7 The circuit symbol for a transformer.

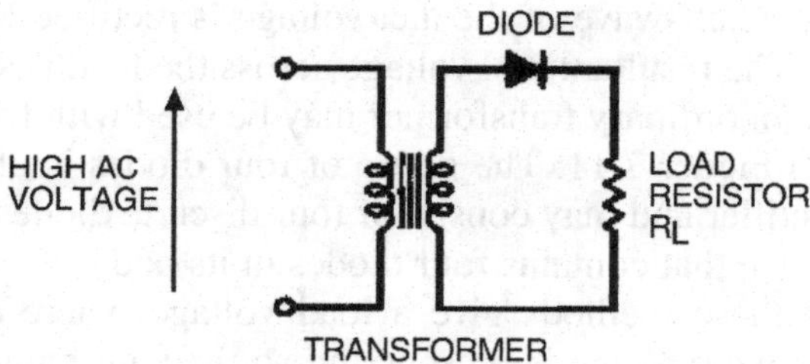

FIGURE 7.8 A simple output rectifier circuit using a diode.

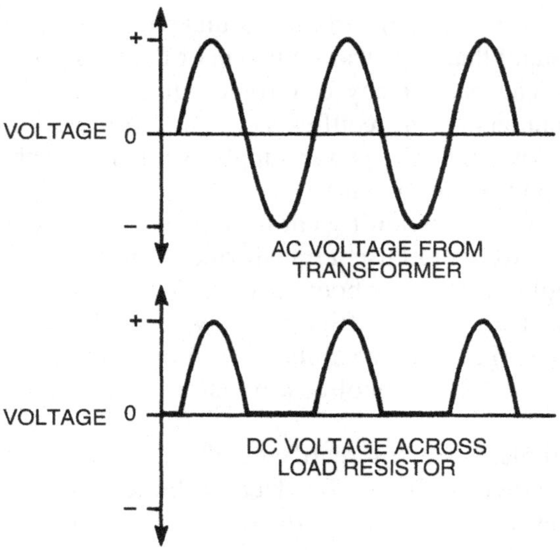

FIGURE 7.9 Waveforms showing the output from the transformers, and the rectified d.c. voltage across the load resistor.

each wave or cycle of a.c. voltage from the transformer, only the positive half gets through the diode to the resistor. For this reason, the type of rectification shown by the circuit in Figure 7.8 is known as half-wave rectification.

It would obviously give a much steadier d.c. voltage if both half-waves of the a.c. voltage could pass to the load. We can do this in two ways. First, by using a modified transformer with a center tap to the output or secondary coil and two diodes as in Figure 7.10. The center tap of the transformer gives a reference voltage to the load, about which one of the two ends of the coil must always have a positive voltage (i.e. if one end is positive the other is negative, if one end is negative the other must be positive), so each half-wave of the a.c. voltage is rectified and passed to the load. The resultant d.c. voltage across the load is shown.

Second, an ordinary transformer may be used with four diodes, as shown in Figure 7.11. The group of four diodes is often called a bridge rectifier and may consist of four discrete diodes or can be a single device that contains four diodes in its body.

Both of these methods give a load voltage where each half-wave of the a.c. voltage is present and so they are known as full-wave rectification.

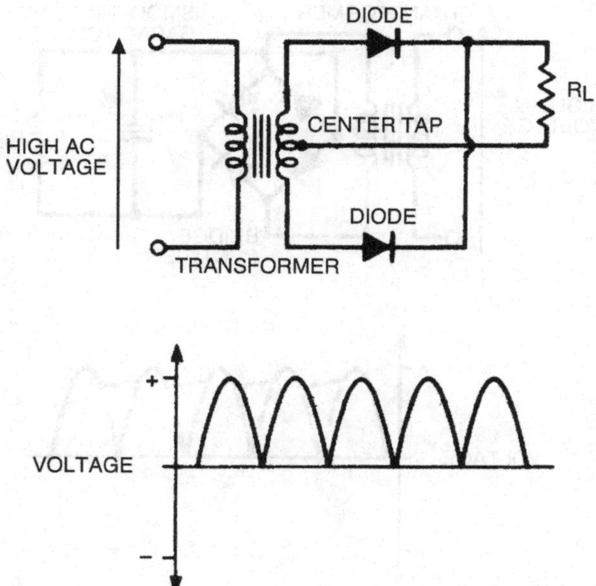

FIGURE 7.10 A more sophisticated rectifier arrangement using two diodes and a transformer center tap.

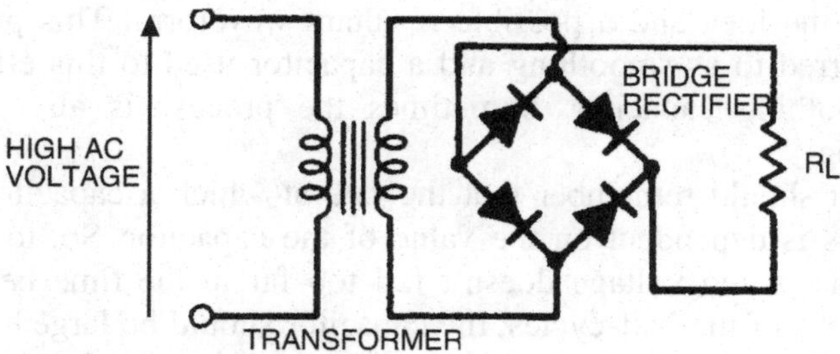

FIGURE 7.11 A familiar rectifier arrangement: four diodes as a bridge rectifier.

FILTER TIPS

Although we've managed to obtain a full-wave rectified d.c. load voltage, we still have the problem that this voltage is not too steady (ideally we would like a fixed d.c. voltage that doesn't vary at all). We can reduce the up-and-down variability of the waves by adding a capacitor to the circuit output. If you remember, a capacitor stores charges – so we can use it to average out the variation in level of the full-wave rectified d.c. voltage. Figure 7.12

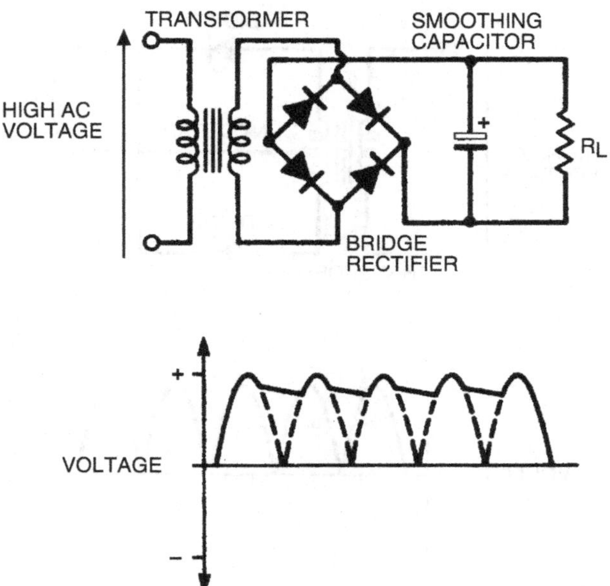

FIGURE 7.12 Leveling the rectifier d.c. with the help of a capacitor: the process is known as smoothing.

shows the idea and a possible resultant waveform. This process is referred to as smoothing and a capacitor used to this effect is a smoothing capacitor. Sometimes the process is also called filtering.

You should remember that the rate at which a capacitor discharges is dependent on the value of the capacitor. So, to make sure the stored voltage doesn't fall too far in the time between the peaks of the half-cycles, the capacitor should be large enough (typically of the value of thousands of microfarads) to store enough charge to prevent this happening. Nevertheless, a variation in voltage will always occur, and the extent of this variation is known as the ripple voltage, shown in Figure 7.13. Ripple voltages of the order of a volt or so are common, superimposed on the required d.c. voltage, for this type of circuit.

STABILITY BUILT IN

The power supplies we've seen so far are simple but they do have the drawback that output voltage is never exact – ripple voltage and, to a large extent, load current requirements mean that a

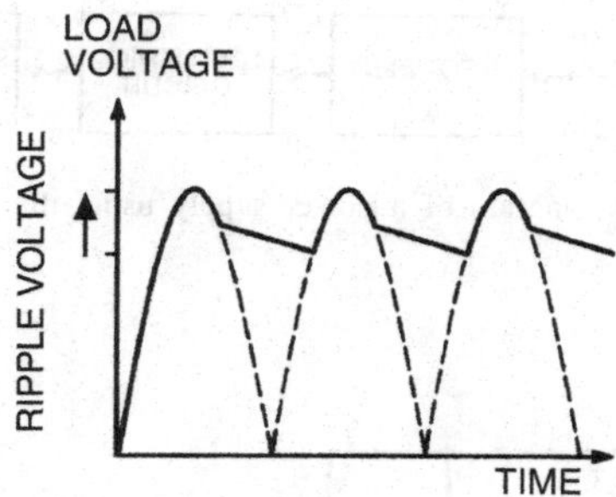

FIGURE 7.13 The extent to which the d.c. is not exactly linear is known as the ripple voltage.

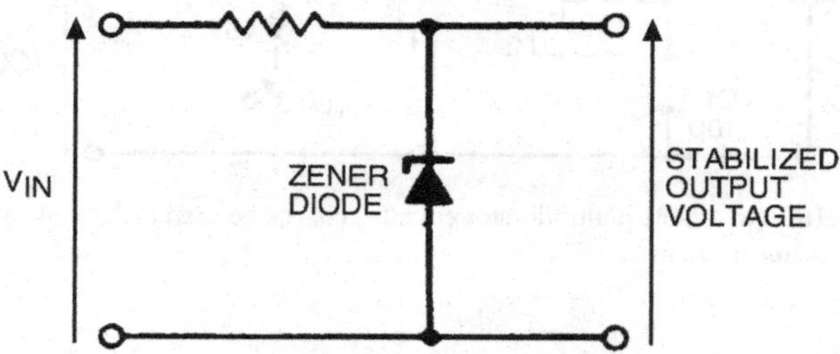

FIGURE 7.14 A simple zener circuit that can be used with a smoothing circuit to give a greatly reduced ripple voltage.

variation in voltage will always occur. In many practical applications such supplies are adequate, but some applications require a much more stable power supply voltage.

We've already seen a device capable of stabilizing or regulating power supplies – the zener diode. Figure 7.14 shows the simple zener circuit we first saw in the last chapter. You'll remember that the zener diode is reverse biased and maintains a more or less constant voltage across it, even as the input voltage V_{in} changes. If such a zener circuit is used at the output of a smoothed power supply (say, that of Figure 7.12), then the resultant stabilized power supply will have an output voltage that is much more stable, with a much reduced ripple voltage.

Zener stabilizing circuits are suitable when currents of no more than about 50 mA or so are required from the power supply. Above this it's more usual to build power supplies using stabilizing integrated circuits (ICs), specially made for the purpose. Such ICs, commonly called voltage regulators, have diodes and

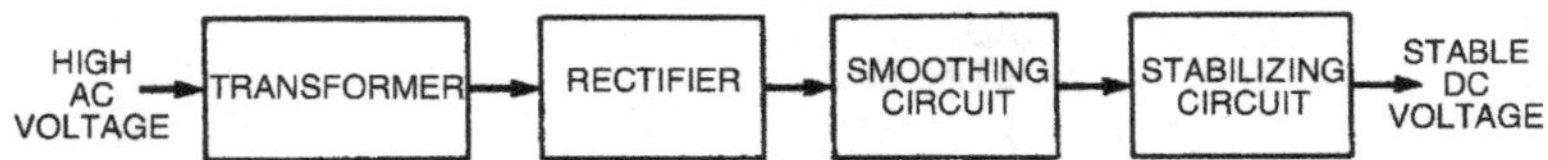

FIGURE 7.15 A block diagram of a power supply using the circuit stages we have described.

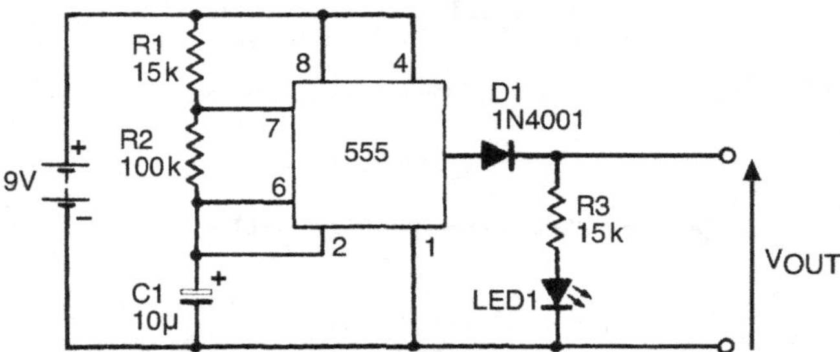

FIGURE 7.16 An astable multi-vibrator circuit. This can be used to demonstrate some of the principles under discussion.

other semiconductors within their bodies that provide the stabilizing stage of the power supply. Voltage regulators give an accurate and constant output voltage with extremely small ripple voltages, even with large variations in load current and input voltage. ICs are produced that can provide load current up to about 5 A.

The power supply principle is summarized in Figure 7.15 in block diagram form. From a 230 V a.c. input voltage, a stabilized d.c. output voltage is produced. This is efficiently done only with the use of diodes in the rectification and stabilization stages.

PRACTICALLY THERE

Figure 7.16 shows a circuit we've already seen and built before: a 555-based astable multi-vibrator. We're going to use it to demonstrate the actions of some of the principles we've seen so far. Although the output of the astable multi-vibrator is a square wave, we're going to imagine that it is a sine wave such as that from a 230 V a.c. mains-powered transformer. Figure 7.17(a) shows the square-wave output, and if you stretch your imagination a bit (careful now!), rounding off the tops and bottoms of the waveform, you can approximate it to an alternating sine wave.

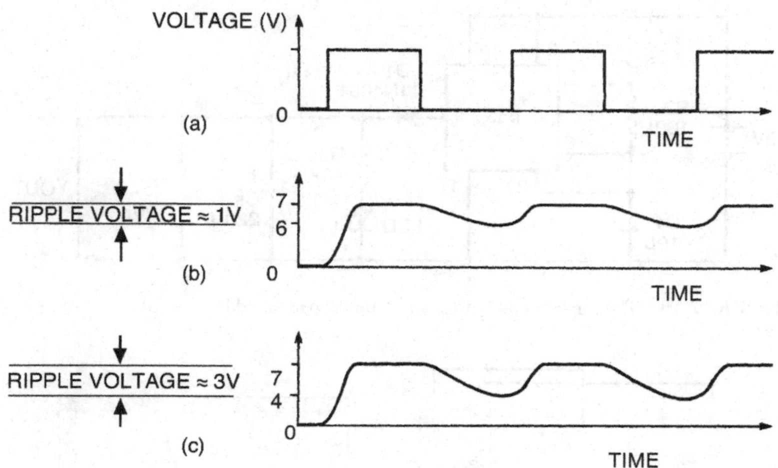

FIGURE 7.17 The square-wave output from the circuit in Figure 7.16 (a), experimentally modified by a 100 kΩ resistor (b), and a 10 kΩ resistor (c).

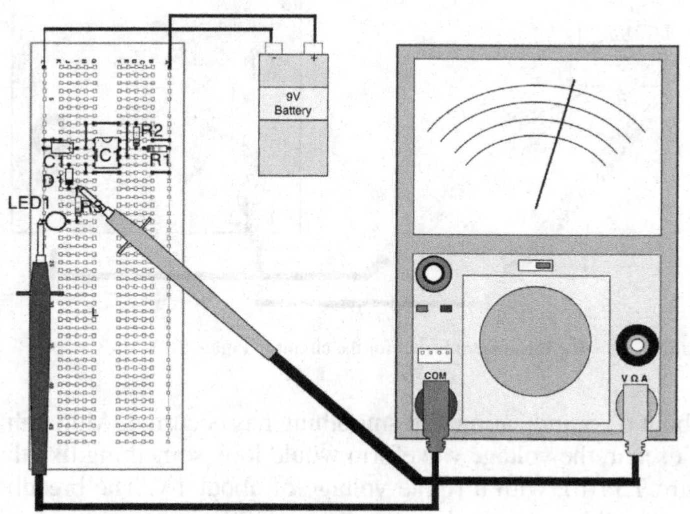

FIGURE 7.18 The breadboard layout of the circuit in Figure 7.16.

Build the circuit, as shown in the breadboard layout of Figure 7.18, and measure the output voltage. It should switch between 0 V and about 8 V.

Now let's add a smoothing capacitor C_s and load resistor R_L to the output of our imaginary rectified power supply, as in Figure 7.19. The output voltage of the smoothing stage is now fairly steady

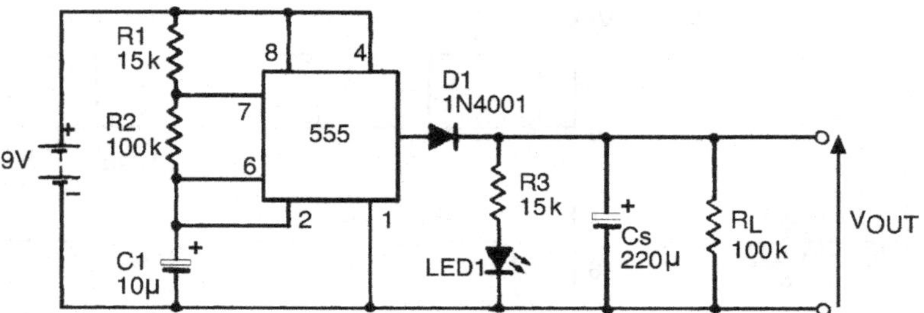

FIGURE 7.19 The same circuit with a smoothing capacitor added.

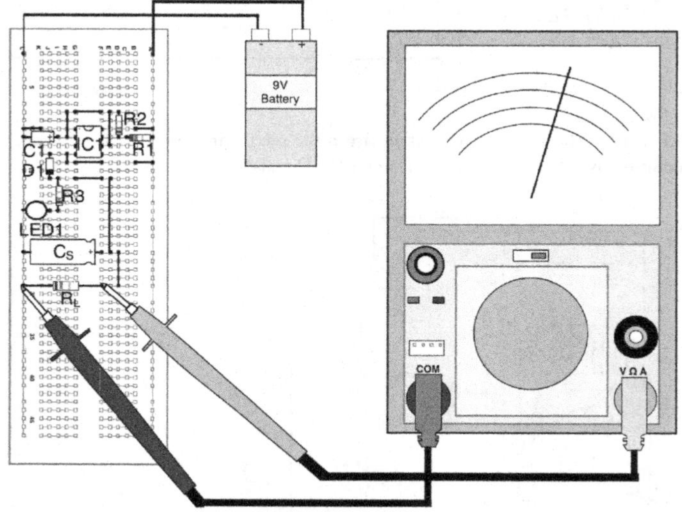

FIGURE 7.20 The breadboard layout for the circuit in Figure 7.19.

at about 6V5, indicating that smoothing has occurred. Although you can't see it, the voltage waveform would look something like that in Figure 7.17(b), with a ripple voltage of about 1 V. The breadboard layout for the circuit is shown in Figure 7.20.

We can show what happens when load current increases by putting a lower value load resistor in the circuit – take out the 100 kΩ resistor and put a 10 kΩ resistor in its place. The ripple voltage now increases, causing greater changes in the d.c. output level, and the whole waveform is shown as Figure 7.17(c).

That's all for this chapter. In the next chapter we'll move on to another semiconductor – possibly the most important semiconductor ever – the transistor. Try the quiz that follows to see what you've learned here.

QUIZ

Answers at the end of the book.

1. Resistor characteristic curves are:
   a. Nearly always linear
   b. Always exponential
   c. Nearly always exponential
   d. Rarely drawn
   e. b and d
   f. All of these.
2. Diodes are:
   a. Ohmic
   b. Non-ohmic
   c. Linear
   d. Non-linear
   e. a and c
   f. b and d
   g. None of these.
3. The voltage across the diode whose characteristic curve is shown in Figure 7.1 is 0V7. What is the current through it?:
   a. 60 mA
   b. About 700 mV
   c. About 6 mA
   d. It is impossible to say, because diodes are non-ohmic
   e. None of these.
4. Mains voltages are dangerous because:
   a. They are a.c.
   b. They are d.c.
   c. They are high
   d. They are not stabilized
   e. None of these.
5. A single diode cannot be used to full-wave rectify an a.c. voltage: true or false?
6. For full-wave rectification of an a.c. voltage:
   a. A bridge rectifier is always needed
   b. As few as two diodes can be used
   c. An IC voltage regulator is essential
   d. a and c
   e. All of these
   f. None of these.

Chapter 8

Transistors

In previous chapters we've looked at semiconductors, spending some time on diodes (the simplest semiconductor device) and taking brief glimpses at integrated circuits. It's now the turn of the transistor to come under the magnifying glass – perhaps the most important semiconductor of all.

You don't need many components in this chapter, just a handful of resistors:

- 1 × 100 kΩ
- 1 × 220 kΩ
- 1 × 47 kΩ miniature horizontal preset

and, of course, one transistor.

The type of transistor you need is shown in Photo 8.1 and is identified as a 2N3053. You'll see that the transistor has three terminals called, incidentally, base, collector, and emitter (often shortened to B, C, and E). When you use transistors in electronic circuits it is essential that these three terminals are the right way round. The 2N3053 transistor terminals are identified by holding the transistor with its terminals pointing towards you from the body and comparing the transistor's underneath with the diagram in Figure 8.1. The terminal closest to the tab on the body is the emitter, then in a clockwise direction are the base and the collector terminals.

Other transistor varieties may have different body types, so it's important to check with reference books or manufacturer's data regarding the transistor terminals before use. All 2N3053 transistors, however, have the same body type – known as a TO-5 body – and follow the diagram in Figure 8.1.

In the previous two chapters, we've looked closely at diodes – the simplest of the group of components known as semiconductors. The many different types of diodes are all formed by combining doped layers of semiconductor material at a junction.

PHOTO 8.1 A transistor next to a UK 50 pence piece.

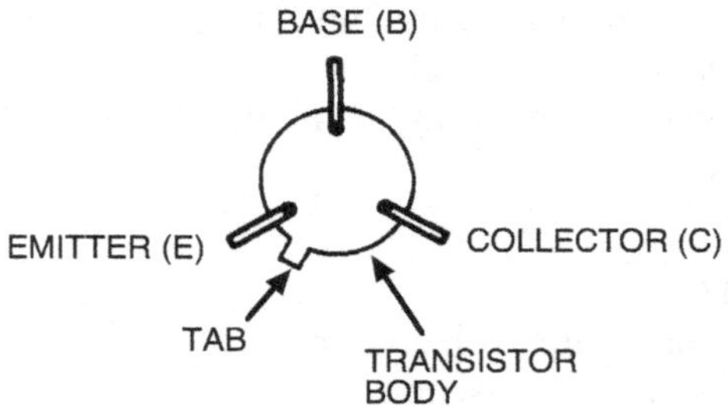

FIGURE 8.1 How to recognize the location of the base, emitter, and collector connections on a common transistor body.

The PN junction (as one layer is N-type semiconductor material and the other layer is P-type) forms the basis of all other semiconductor-based electronic components. The transistor – the component we're going to look at now – is, in fact, made of two PN junctions, back to back. Figure 8.2 shows how we may simply consider a transistor as being two back-to-back diodes, and we can verify this using our multimeter to check resistances between the three terminals of a transistor: the base, emitter, and collector.

To do the experiment, put a transistor into your breadboard, then use the meter to test the resistance between transistor terminals. Using our brains here (I know it's hard) we can work out that if there are three terminals there must be six different ways the meter leads can connect to pairs of the terminals. Table 8.1 lists all six ways, but the results column is left blank for you to fill in as you perform the experiment. It's not necessary for you to measure the exact resistance obtained between terminals, it's sufficient just to find out if the resistance is high or low.

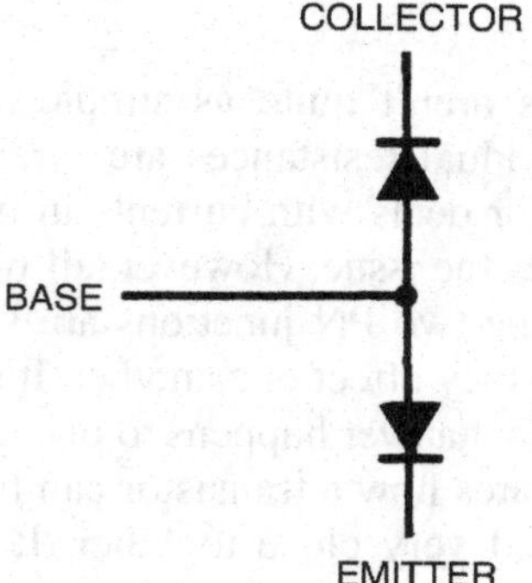

FIGURE 8.2 A symbol for a transistor considered as two diodes, back to back.

TABLE 8.1 Six Different Ways to Connect the Meter Leads

Black Meter Lead	Red Meter Lead	Result
Base	Emitter	
Emitter	Base	
Base	Collector	
Collector	Base	
Emitter	Collector	
Collector	Emitter	

Take Note

One final point before you start – the casing of the transistor body (sometimes called the can) is metal and therefore conducts. When you touch the meter test probes to the transistor terminals make sure, therefore, they don't touch the can. It's all too easy to do without realizing you've got a short circuit that, of course, gives an incorrect result. The can of a TO-5 bodied transistor such as the 2N3053 is also electrically connected to the transistor's collector, so be careful!

Your results should show that low resistances occur in only two cases, indicating forward current flow between base and emitter, and base and collector. This corresponds, as we would expect, to the diagram of Figure 8.2.

VERY CLOSE

Unfortunately things aren't quite as simple as that in electronic circuits, where individual resistances are rarely considered alone. The real-life transistor deals with currents in more than one direction and this confuses the issue. However, all we need to know here (thankfully) is that the two PN junctions are very close together – so close that, in fact, they affect one another. It's almost as if they're Siamese twins – and whatever happens to one affects the other.

Figure 8.3 illustrates how a transistor can be built up, from two PN junctions situated very close together. It's really only a thin layer of P-type semiconductor material (only a few hundred or so atoms thick) between two thicker layers of N-type semiconductor material. Now let's connect this transistor arrangement between a voltage supply, so that collector is positive and emitter is zero, as in Figure 8.4.

From what we know so far, nothing can happen and no current can flow from collector to emitter because between these two terminals two back-to-back PN junctions lie. One of these junctions is reverse biased and so, like a reverse-biased diode, cannot conduct. So, what's the point of it all?

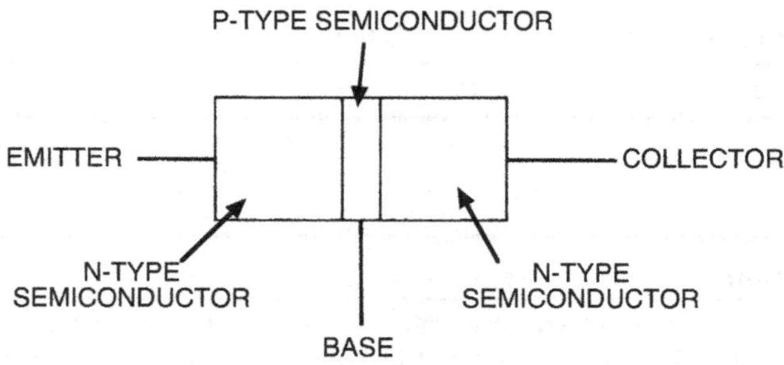

FIGURE 8.3 A narrow P-area gives two PN junctions very close together.

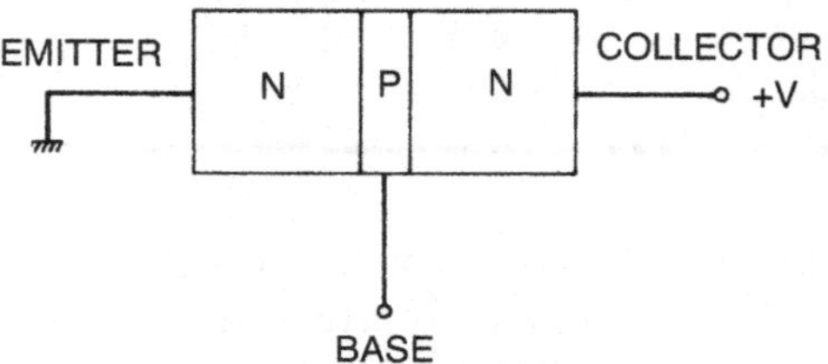

FIGURE 8.4 Connected up, the emitter is at 0 V and the collector is positive.

Well, as mentioned earlier, what happens in one of the two PN junctions of the transistor affects the other. Let's say, for example, that we start the lower PN junction (between base and emitter) conducting by raising the base voltage so that the base-to-emitter voltage is above the transition voltage of the junction (say, 0V6 if the transistor is a silicon variety). Figure 8.5 shows this situation. Now, the lower junction is flooded with charge carriers and because both junctions are very close together, these charge carriers also allow current flow from collector to emitter, as shown in Figure 8.6.

So, to summarize, a current will flow from collector to emitter of the transistor when the lower junction is forward biased by a small base-to-emitter voltage. When the base-to-emitter voltage is removed the collector-to-emitter current will stop.

We can build a circuit to see if this is true, as shown in Figure 8.7. Note the transistor circuit symbol. Figure 8.8 shows the breadboard layout. From the circuit you'll see that we're measuring the transistor's collector-to-emitter current (commonly shortened to just collector current) when the base-to-emitter junction is first connected to zero volts and to all intents and purposes is reverse biased, then when the base is connected to positive and the base-to-emitter junction is forward biased.

Now do the experiment and see what happens. You should find that when the base is connected to zero via the 100 kΩ resistor

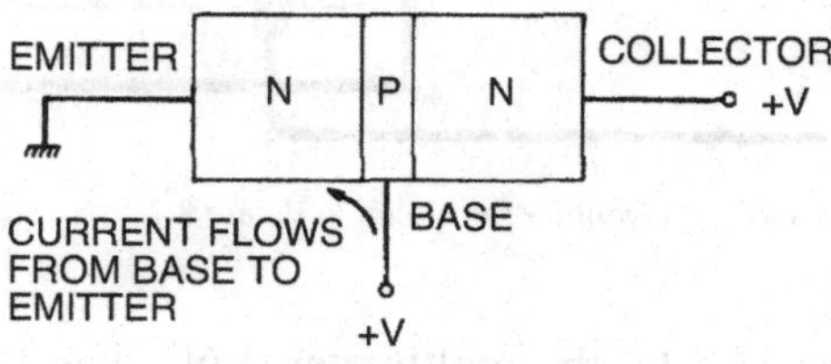

FIGURE 8.5 If the voltage at the base is raised, current flows from base to emitter.

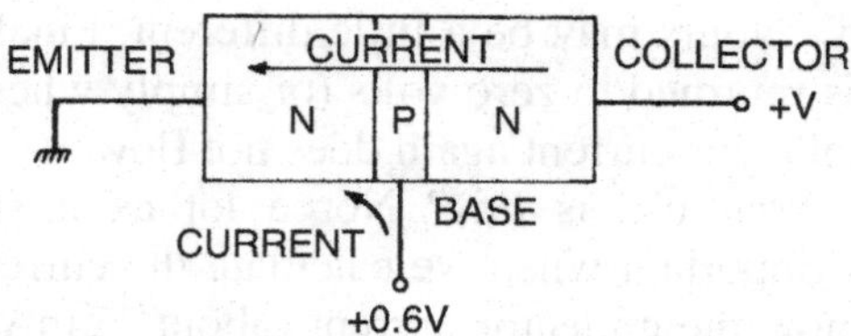

FIGURE 8.6 Charge carriers accumulating around the lower junction allow current to flow from the collector to the emitter.

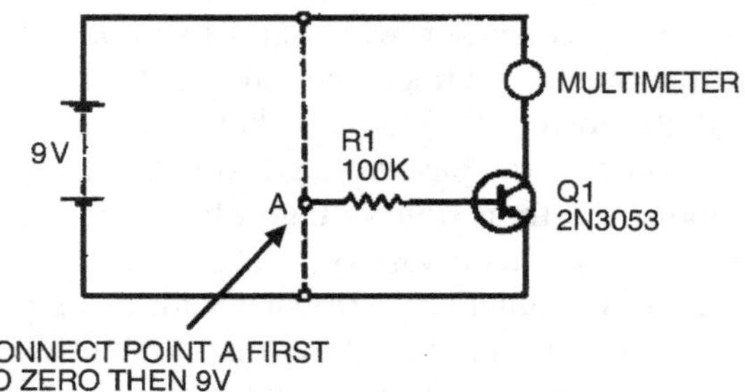

FIGURE 8.7 An experimental circuit to test what we have described so far.

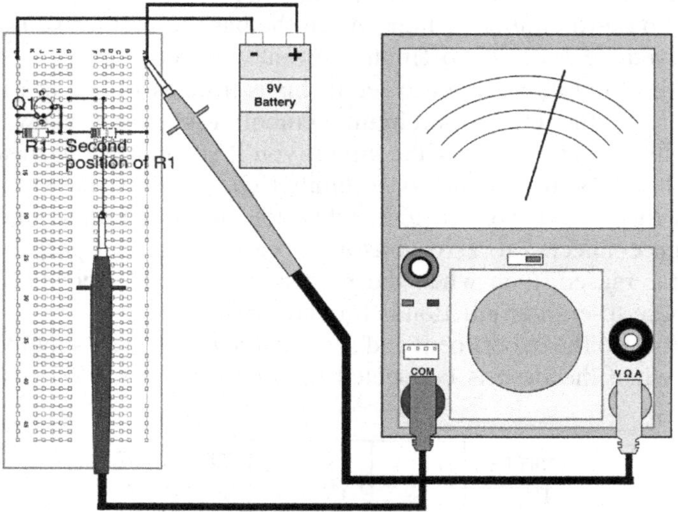

FIGURE 8.8 A breadboard layout for the circuit in Figure 8.7.

nothing is measured by the multimeter. But when the base resistor is connected to positive, the multimeter shows a collector current flow. In our experiment a collector current of about 12 mA was measured – yours may be a little different. Finally, when the base resistor is returned to zero volts (or simply when it's disconnected!) the collector current again does not flow.

So what? What use is this? Not a lot as it stands, but it becomes very important when we calculate the currents involved. We already know the collector current (about 12 mA in our case) but what about the base-to-emitter current (shortened to base current)? The best way to find this is not by measurement (the meter

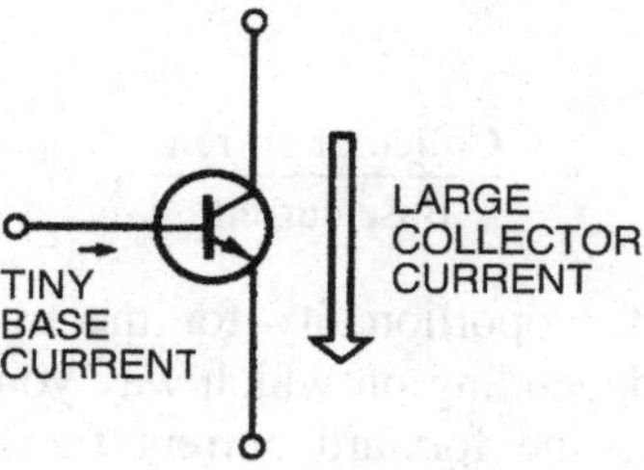

FIGURE 8.9 Important: a very small base current can control a large collector current.

itself would affect the transistor's operation!) but by calculation. We know the transistor's base-to-emitter voltage (shortened to base voltage) and we know the supply voltage. From these we can calculate the voltage across the resistor, and from Ohm's law we can therefore calculate the current through the resistor. And the current through the resistor must be the base current.

The resistor voltage is:

$$9 - 0.7 = 8.3\ \text{V}$$

So, from Ohm's law the current is:

$$I = \frac{V}{R} = \frac{8.3}{100{,}000} = 83\ \mu\text{A}$$

Not a lot!

Now we can begin to see the importance of the transistor. A tiny base current can turn on or off a quite large collector current. This is illustrated in Figure 8.9 and is of vital importance – so remember it!

In effect, the transistor is a current amplifier. No matter how small the base current is, the collector current will be much larger. The collector current is, in fact, directly proportional to the base current. Double the base current and you double the collector current. Halve the base current and the collector current is likewise halved. It's this fact that the transistor may be used as a controlling element (the base current controlling the collector current) that makes it the most important component in the electronics world.

The ratio

$$\frac{\text{Collector current}}{\text{Base current}}$$

gives a constant of proportionality for the transistor, which can have many names depending on which way you butter your bread. Officially it's called the forward current transfer ratio, common emitter, but as that's quite a mouthful it's often just called the transistor's current gain (seems sensible!). You can sometimes shorten this even further if you wish, to the symbols h_{fe} or β. In manufacturers' data sheets for transistors the current gain is normally just given the symbol h_{fe} (which if you're interested stands for Hybrid parameter, Forward, common Emitter – are you any wiser?). However, we'll generally just stick to the term current gain.

We can work out the current gain of a transistor by measuring the collector current, and calculating the base current as we did earlier, and dividing one by the other. For example, the current gain of the transistor we used is:

$$\frac{12 \times 10^{-3}}{83 \times 10^{-6}} = 145$$

Yours may be a bit different. Manufacturers will quote typical values of current gain in their data sheets – individual transistors' current gains will be somewhere around this value, and may not be exact at all. It really doesn't matter too much. The transistor we use here, the 2N3053, is a fairly common general-purpose transistor. High-power transistors may have current gains more in the region of about 10, while some modern transistors for use in high-frequency circuits such as radio may have current gains around 1000 or so.

NPN

The 2N3053 transistor is known as an NPN transistor because of the fact that a thin layer of P-type semiconductor material is sandwiched between two layers of N-type semiconductor material. The construction and circuit symbol of an NPN transistor are shown in Figure 8.10(a). You may have worked out that there is another way to sandwich one type of semiconductor material

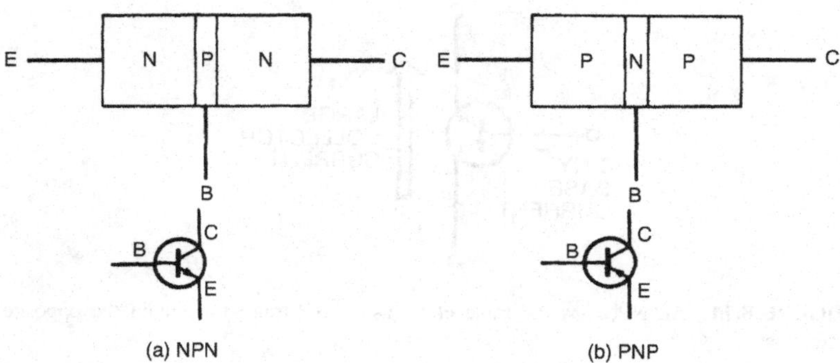

FIGURE 8.10 The internal construction and circuit symbols for NPN and PNP transistors.

between two others – and if so you would also have worked out that such a transistor would be called a PNP transistor. Figure 8.10(b) shows a PNP transistor construction and its circuit symbol. The only difference in the circuit symbols of both types is that the arrow on an NPN transistor's emitter points out and the arrow on a PNP transistor's emitter points in.

The emitter arrow of either symbol indicates direction of base current and collector current flow. So from the circuit symbols we can work out that base current in the NPN transistor flows from base to emitter, while in the PNP transistor it flows from emitter to base. Likewise, collector current flow in the NPN transistor is from collector to emitter and from emitter to collector in the PNP transistor.

Knowing this and comparing the PNP construction to that of the NPN transistor we can further work out that a tiny emitter-to-base current (still called the base current, incidentally) will cause a much larger emitter-to-collector current (still called the collector current). This is illustrated in Figure 8.11. The ratio of collector current to base current of a PNP transistor is still the current gain. In fact, apart from the different directions of currents, a PNP transistor functions identically to an NPN transistor. As we've started our look at transistors with the use of an NPN transistor, however, we'll finish it the same way.

USING TRANSISTORS

We've seen how transistors work but we don't yet know how they can be used. After all, there are millions and millions of

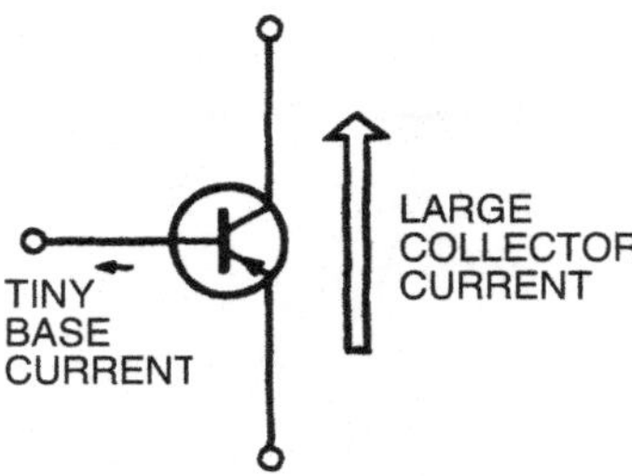

FIGURE 8.11 An NPN has the same effect as a PNP transistor, but in the opposite direction.

transistors around in the world today – you'd be forgiven for thinking that there must be hundreds if not thousands of ways that a transistor may be used. Well, you'd be forgiven, but you'd still be wrong!

In fact, to prove the point we're now going to go through every possible way a transistor can be made to operate, and as this chapter isn't five miles thick you've probably realized that there aren't many ways a transistor may be used at all. Incredibly, at the last count, the number of ways a transistor may be used is – dah, da, dah, dah, dah, da, dah, dah – two!

Hint

Yes, that's right, only two basic uses of a transistor exist, and every transistorized circuit, every piece of electronic equipment, every television, every radio, every computer, every digital watch, and so on, contains transistors in one form or another which do only one of two things.

We've already seen the first of these two uses – an electronic switch, where a tiny base current turns on a comparatively large collector current. This may appear insignificant in itself, but if you consider that the collector current of one transistor may be used as the base current of a following transistor or transistors, then you should be able to imagine an enormous number of transistors inside one appliance, all switching and hence controlling the appliance's operation.

Incredible? Science fiction? Well, let me tell you that the appliance described here in extremely simple terms already exists, in millions. We call the appliance a computer. And every computer contains thousands if not millions of transistor electronic switches.

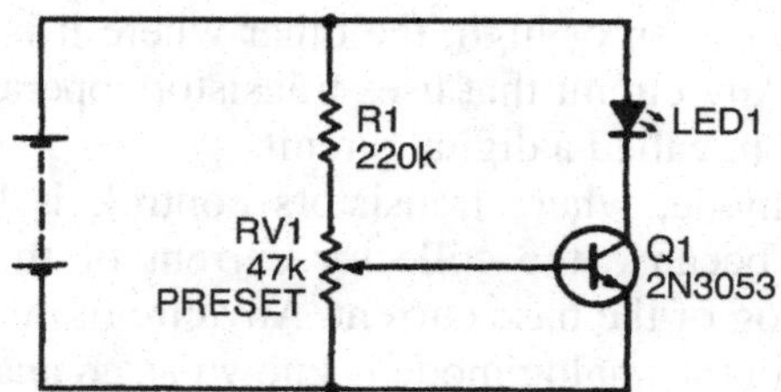

FIGURE 8.12 A circuit to test a transistor with a variable base current.

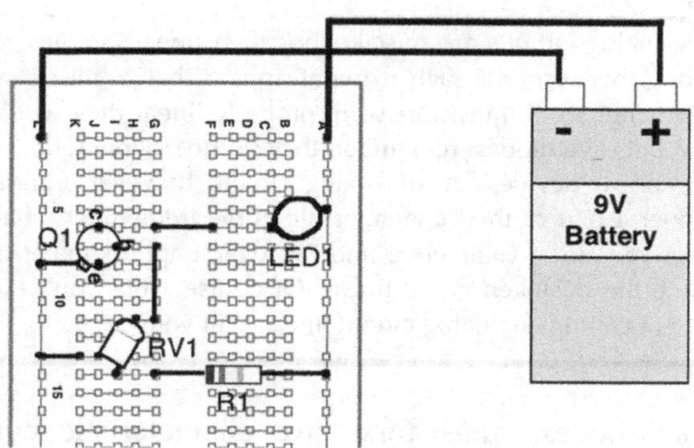

FIGURE 8.13 A breadboard layout for the circuit in Figure 8.12. The preset should be turned slowly with a fine screwdriver.

We can see the second use of a transistor in the circuit of Figure 8.12. From this you should be able to see that we're going to use a variable resistor to provide a variable base current for the transistor in the circuit. The breadboard layout of the circuit is shown in Figure 8.13. Before you connect the battery, make sure the preset variable resistor is turned fully anticlockwise.

Now, connect the battery and slowly (with a small screwdriver) turn the preset clockwise. Gradually, as you turn the preset, the LED should light up: dim at first, then brighter, then fully bright. What you've built is a very simple lamp dimmer.

So, the other use of a transistor is as a variable control element. By controlling the transistor's tiny base current we can control the much larger collector current, which can be the driving current of, say, an LED, an ordinary lamp, a motor, a loudspeaker – in fact, just about anything that is variable.

These two operational modes of transistors have been given names. The first, as it switches between two states, one where the collector current is on or high, the other where it is off or low, is called digital. Any circuit that uses transistors operating in digital mode is therefore called a digital circuit.

The other mode, where transistors control, is known as the analog mode, because the collector current of the transistor is simply an analog of the base current. Any circuit that uses transistors operating in the analog mode is known as an analog circuit.

Take Note

Sometimes analog circuits are mistakenly called linear circuits. However, this is wrong, because although it might appear that a linear law is followed, this is not so. If transistors were properly linear they would follow Ohm's law, but (like diodes) remember, they do not follow Ohm's law and so are non-linear devices. In an analog circuit, however, transistors are operated over a part of their characteristic curve (remember what a characteristic curve is from Chapters 6 and 7?), which approximates a straight line – hence the mistaken name linear. Of course, you won't ever make the mistake of calling an analog circuit linear, will you?

Yes, like diodes, transistors have characteristic curves too, but as they have three terminals they have correspondingly more curves.

QUIZ

Answers at the end of the book.

1. In the circuit symbol for an NPN transistor:
   - **a.** The arrow points out of the base
   - **b.** The arrow points into the base
   - **c.** The arrow points out of the collector
   - **d.** The arrow points into the collector
   - **e.** b and c
   - **f.** All of these
   - **g.** None of these.
2. In an NPN transistor:
   - **a.** A small base current causes a large collector current to flow
   - **b.** A small collector current causes a large base current to flow
   - **c.** A small emitter current causes nothing to flow
   - **d.** Nothing happens
   - **e.** None of these
   - **f.** All of these.

3. If an NPN transistor has a current gain of 100, and a base current of 1 mA, its collector current will be:
  a. 1 mA
  b. 1 μA
  c. 10 μA
  d. 100 μA
  e. Bigger
  f. None of these.

4. The current gain of a transistor:
  a. Is equal to the collector current divided by the emitter current
  b. Is equal to the collector current divided by the base current
  c. Is equal to the base current divided by the collector current
  d. a and c
  e. All of these
  f. None of these.

5. The current gain of a transistor has units of:
  a. Amps
  b. Milliamps
  c. Volts
  d. This is a trick question, it has no units
  e. None of these.

Chapter 9

Analog Integrated Circuits

Well, the last chapter was pretty well jam-packed with information about transistors. If you remember, we saw that transistors are very important electronic components. They may be used in one of only two ways: in digital circuits or in analog circuits – although many, many types of digital and analog circuits exist.

In terms of importance, transistors are the tops. They're far more important than, say, resistors or capacitors, although in most circuits they rely on the other components to help them perform the desired functions. They're the first electronic components we've come across that are active: they actually control the flow of electrons through themselves, to perform their function – resistors and capacitors haven't got this facility, they are passive and current merely flows or doesn't flow through them.

Transistors, being active, can control current so that they can be turned into amplifiers or switches depending on the circuit. There's nothing magical about this, mind you, we're not gaining something for nothing! In order that, say, a transistor can amplify a small signal into a large one, energy has to be added in the form of electricity from a power supply. The transistor merely controls the energy available from this power supply, creating the amplification effect.

They're important for another reason, however. They are small! They can be made by mass-production techniques, are almost as small as you can imagine, and certainly many times smaller than you could see with the naked eye. This in itself is no big deal but imagine trying to solder a transistor into a circuit that was so small you couldn't even see it! Transistors of the types we've seen so far are purposefully made as large as they are just so we can handle them.

INTEGRATED CIRCUITS

But the big advantage of transistors, especially tiny transistors, is that many of them may be integrated into a single circuit, which is itself still very small. To give you an idea of what we're talking about, modern integrated circuits (well, what else would you call them?) have been made with over a quarter of a million transistors, all fitting on to one small silicon chip only about 40 square millimeters or so! Now this sort of integration represents the ultimate in human achievement, remember, but processes are being improved every year and integrated circuits with only hundreds of thousands of integrated transistors are now commonplace. Because of this it stands to reason that we have to come to grips with integrated circuits, they're here to stay and there's no point in being shy. So the final chapters of this epic saga are devoted to integrated circuits.

The integrated circuit that we are going to play with in this chapter is an analog integrated circuit (let's cut all these long words out and call it by its abbreviation – IC). Just as there are analog and digital circuits that transistors are used in, so there are analog and digital ICs. This IC just happens to be analog, but there are digital ones too – the 555 (which we've already seen) is actually an example. In IC terms the IC in this chapter is a comparatively simple IC, having only around 20 transistors in all; nevertheless, it is an extremely versatile IC and is probably the most commonly used IC of all time.

Generally it's called the 741, referring to the numbers that are printed on the top of the IC body (see Photo 9.1). If you look at your IC you'll see that it also has some letters associated with the number 741. These letters, e.g. LM, SN, μ, MC, refer to the manufacturer of the device and bear no relevance to the internal circuit, which is the same in all cases. An SN741 is identical in

PHOTO 9.1 Our subject: a 741 op-amp.

performance to a μ741. Other letters printed on the top of the IC refer to other things such as batch number, date of manufacture, etc. Whatever is printed on the top of the IC, as long as it is a 741 there's normally no problem and it'll work in all the circuits we're going to look at now.

You'll notice that the 741 is identical in physical appearance to the other IC we've already looked at, the 555. Like the 555, the 741 is housed in an eight-pin DIL (dual-in-line) body. But although the shape's the same, the internal circuit isn't. Other versions of the 741 exist, in different body styles, but the eight-pin DIL version is by far the most popular.

The 741's internal circuit is shown in Figure 9.1. Its circuit symbol is shown in Figure 9.2. From this you can see that the

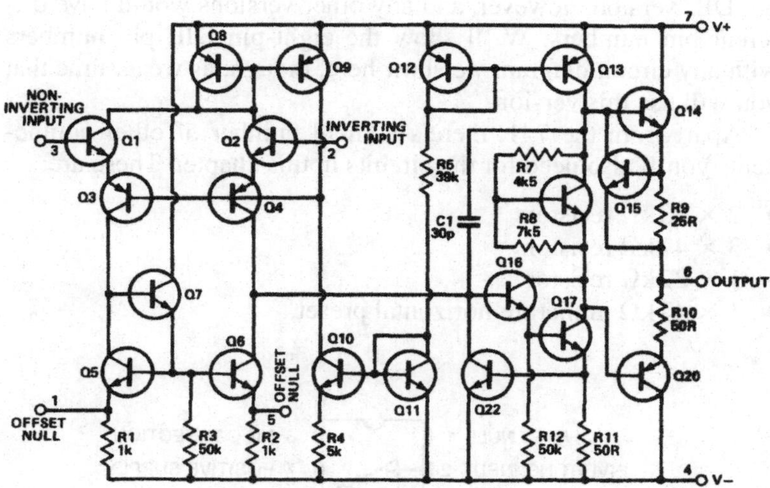

FIGURE 9.1 The internal circuit of the 741: a mere 22 transistors.

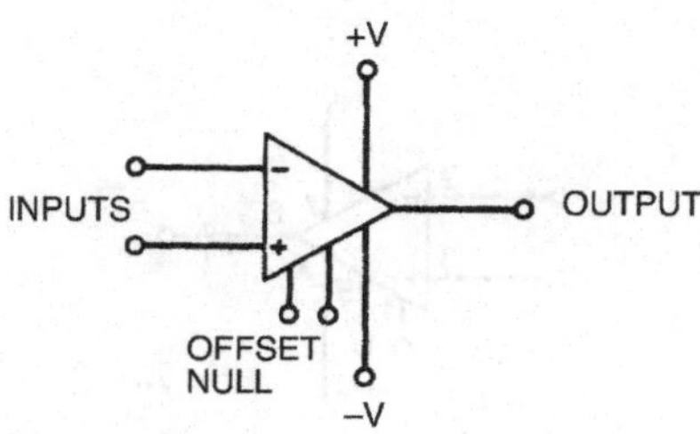

FIGURE 9.2 The circuit symbol for a 741.

circuit has two inputs, marked + and −. Technically speaking these are called the non-inverting and inverting inputs. The circuit has one output and two inputs for power supply connection. There are also two other connectors to the circuit, known as offset null connections; by controlling the voltages of these inputs, we can control the voltage level of the output – you'll find out how these connections work soon. They are not used in every circuit that the 741 is used in.

The internal layout of the eight-pin DIL version of the 741 is shown in Figure 9.3. Now, each input and output of the op-amp is associated with a particular pin of the DIL body. This means that we can redraw the circuit symbol of the 741 as in Figure 9.4, with the corresponding pin numbers at each connection. Note that these pin numbers are only correct when we deal with the eight-pin DIL version, however, and any other versions would have different pin numbers. We'll show the eight-pin DIL pin numbers with any circuit diagram we show here, though, as we assume that you will use this version.

Apart from the 741, there's a small number of other components you'll also need for the circuits in this chapter. These are:

- 2 × 10kΩ resistors
- 3 × 47kΩ resistors
- 1 × 22kΩ resistor
- 1 × 47kΩ miniature horizontal preset.

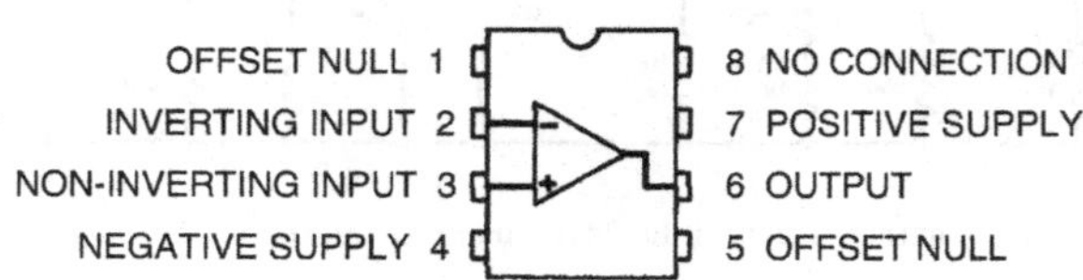

FIGURE 9.3 The internal layout of an eight-pin DIL-packed 741.

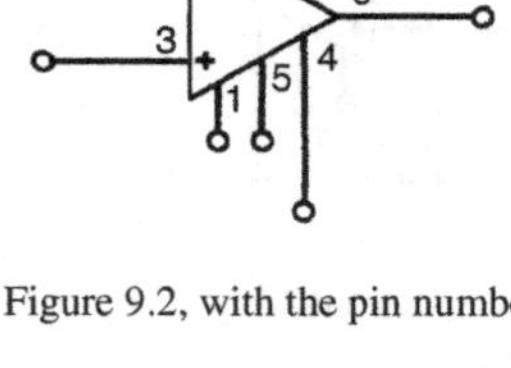

FIGURE 9.4 The symbol in Figure 9.2, with the pin numbers included.

BUILDING ON BLOCKS

The first circuit is shown in Figure 9.5, and from this you will see two 9 V batteries are used as a power supply. This is because the 741 needs what's called a three-rail power supply, i.e. one with a positive rail, a negative rail, and a ground rail 0V. Now you may ask how two batteries can be used to provide this. Well the answer's simple: join them in series. This gives a +18V supply, a 0V supply, and at the mid-point between the batteries, a +9V supply. As voltages are only relative anyway, we can now consider the +9V supply to be the middle ground rail of our three-rail power supply and refer to it as 0V. The +18V supply then becomes the positive rail (+9V) and the 0V supply becomes the negative rail (−9V).

Hint

The 741 is commonly referred to as an operational amplifier – shortened to op-amp. The term comes from the earlier days in electronics when discrete components were built into circuits and used as complete general-purpose units. An op-amp, as its name suggests, is a general-purpose amplifier that is more or less ready to use. Years ago it might have been a large metal box full of valves and wiring. Later it would have been a circuit board on which the circuit was built with transistors. But nowadays nobody uses op-amps made with discrete components, because modern technology has produced IC op-amps such as the 741. Given the choice, would you use discrete components to build an op-amp of the complexity of that in Figure 9.1, or would you use a 741?

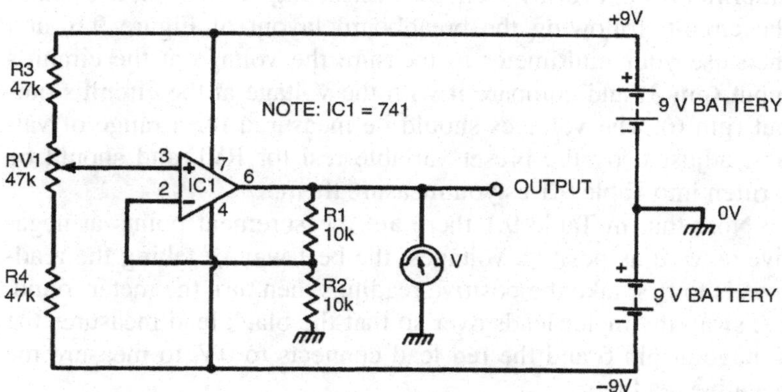

FIGURE 9.5 Our first experimental circuit: a non-inverting amplifier with a three-rail power supply.

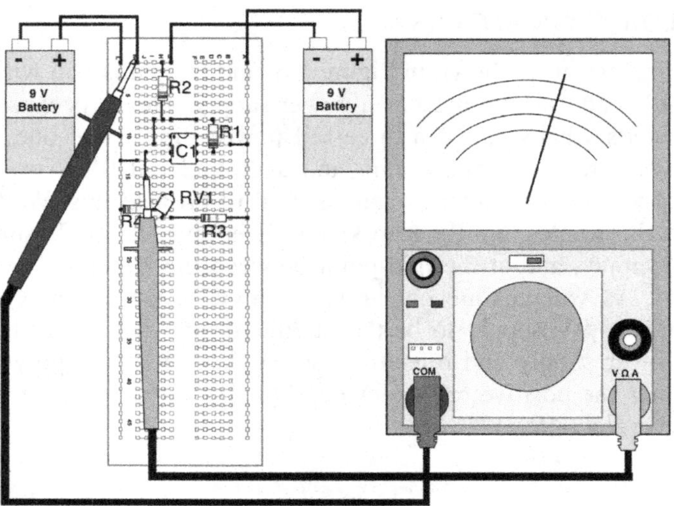

FIGURE 9.6 The breadboard layout for the circuit in Figure 9.5. Note that in this diagram the input voltage is being measured.

Besides the ease of use inherent with ICs, they are also cheaper than their discrete counterparts – the 741 is priced at around 40 pence. Building the circuit using transistors would cost many times this.

As the 741 is an operational amplifier, it makes sense that the first circuit we look at is an amplifying circuit. But what sort of amplifier does it form? Well, the easiest way to find out is to build the circuit, following the breadboard layout of Figure 9.6, and then use your multimeter to measure the voltage at the circuit's input (pin 3) and compare it with the voltage at the circuit's output (pin 6). The voltages should be measured for a range of values, adjusted by the preset variable resistor RVl, and should be written into Table 9.1 as you measure them.

Note that in Table 9.1 there are measurement points at negative as well as positive voltages: the best way of taking the readings is to first take the positive readings then turn the meter round, i.e. swap the meter leads over so that the black lead measures the voltage at pin 6 and the red lead connects to 0V, to measure the negative readings.

The results that were taken in my circuit are given in Table 9.2, and your results should be something like this, although differences

TABLE 9.1 The Results of Your Measurements

Input Voltage	Output Voltage
−2	
−1	
0	
1	
2	

TABLE 9.2 The Results of My Measurements

Input Voltage	Output Voltage
−2	−4.4
−1	−2.1
0	0
1	2.1
2	4.4

may occur. From these results you should see that the circuit amplifies whatever voltage is present at the input by about 2, so that the output voltage is about twice that of the input. This is true whatever the input voltage. Don't worry if your results aren't that close to the ideal ones, it's only an experiment.

This particular circuit is a fairly basic amplifier that simply amplifies the input voltage. For one reason or another, which you'll appreciate soon, it's called a non-inverting amplifier. Its gain is given by the expression:

$$G = \frac{\text{R1}}{\text{R2}} + 1$$

and if you insert the values of resistors R1 and R2 into this expression you will see why the circuit gives an experimental gain of about 2.

As the circuit has a gain that is dependent on the values of resistors R1 and R2, it is a simple matter to adjust the circuit's gain, by changing one or both resistors. Try changing resistor R1 to, say, 22k and see what the gain is (it should be about 3).

Take Note

Theoretically, of course, the circuit could have any gain we desire, just by inserting the right values of resistances. But in practice there are limits. If, say, the circuit's gain was set to be 20, and an input voltage of 1 V was applied, the circuit would attempt to produce an output voltage of 20 V. But that's not possible – the output voltage cannot be greater than the power supply voltage (fairly obvious), which in this case is ±9 V. The answer, you may think, is to wind up the power supply voltage to + and −20 V, and, yes, you can do this to a certain extent, but again there are practical limitations. The 741, for example, allows a maximum power supply voltage of about ±15 V, above which it may be damaged.

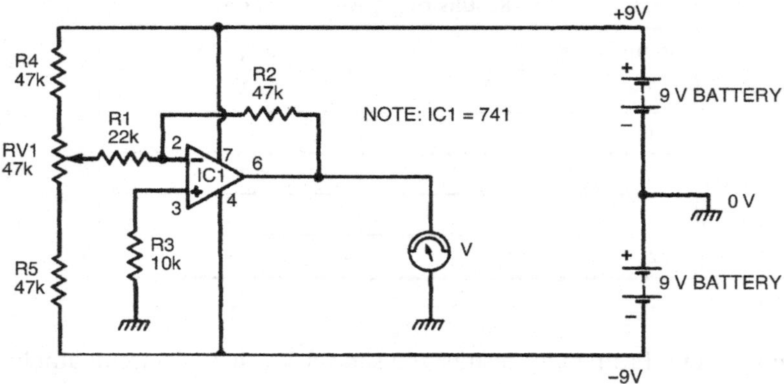

FIGURE 9.7 Our next experimental circuit: an inverting amplifier.

TURNING THINGS UPSIDE DOWN

Earlier, we saw how the op-amp has two inputs, which we called non-inverting and inverting inputs. Now if we used the non-inverting input of the op-amp to make a non-inverting amplifier (Figure 9.5), then it makes sense that the inverting input of the op-amp may be used to make an inverting amplifier, as shown in Figure 9.7.

Build the circuit up on your breadboard, as shown in the breadboard layout of Figure 9.8. In a similar way to the last circuit, now take a number of voltage measurements of input and output voltages, to confirm that this really is an inverting amplifier. Remember that whatever polarity the input voltage is, the output voltage should be the opposite, so you'll have to change round the meter leads to suit.

Tabulate your results in Table 9.3. The results from my circuit are given in Table 9.4, for reference. You'll see that the circuit does invert the voltage, but also amplifies it by 2 at the same time.

Hint

One final point: the input to the op-amp circuit is classed as the connection between resistor R1 and the preset resistor, not the inverting input terminal at pin 2 – you'll get some peculiar results if you try to measure the input voltage at pin 2!

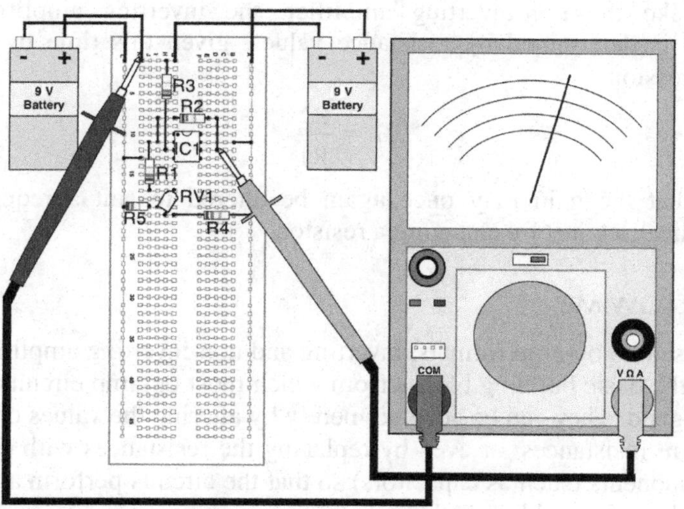

FIGURE 9.8 The breadboard layout for the circuit in Figure 9.7.

TABLE 9.3 The Results of Your Measurements

Input Voltage	Output Voltage
−2	
−1	
0	
1	
2	

TABLE 9.4 The Results of My Measurements

Input Voltage	Output Voltage
−2	4.2
−1	2.1
0	0
1	−2.1
2	−4.2

Like the non-inverting amplifier, the inverting amplifier's gain is determined by resistance values, given this time by the expression:

$$G = \frac{R2}{R1}$$

so that the gain may once again be altered to suit a required application, just by changing a resistor.

FOLLOW ME

These two op-amp formats, inverting and non-inverting amplifiers, are the basic building blocks from which most op-amp circuits are designed. They can be adapted merely by altering the values of the circuit resistances, or even by replacing the resistances with other components (such as capacitors) so that the circuits perform a vast number of possible functions.

As an example we'll look at a couple of circuits, considering how they are designed, compared with the two basic formats. First, we'll consider the circuit in Figure 9.9. At first sight we can see that it is a non-inverting circuit of some description – the input voltage (derived from the wiper of the preset resistor) is applied to the non-inverting input of the op-amp. But in comparison with our non-inverting amplifier of Figure 9.5, there are no resistors around the amplifier; instead, the output voltage is simply looped back to the inverting input. So how does it work and what does it do?

Let's first consider the non-inverting amplifier gain expression:

$$G = \frac{R1}{R2} + 1$$

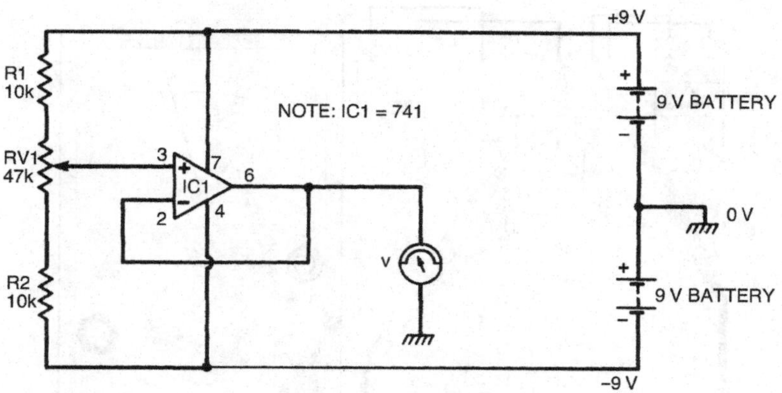

FIGURE 9.9 A voltage follower circuit: the output is equal to the input. This is basically a non-inverting circuit.

and look to see how we can apply it here. Although no resistances are in this circuit, that doesn't mean that we don't need to consider them! Resistor R1 in the circuit of Figure 9.5 can be seen to be between pin 6 of the op-amp (the output) and pin 2 (the inverting input). In the circuit of Figure 9.9, on the other hand, there is a direct connection between pin 6 and pin 2 – in other words, we can imagine that resistor R1 of this circuit is zero.

Similarly, resistor R2 of the circuit in Figure 9.5 lies between pin 2 and 0V. Between pin 2 and 0V of the circuit in Figure 9.9 there is no resistor – in other words, we can imagine the resistance to be infinitely high.

We can put these values of resistances into the expression for gain, which gives us:

$$G = \frac{0}{\infty} + 1 = 1$$

so the circuit has a gain of 1, i.e. whatever voltage is applied at the input is produced at the output. The circuit has a number of common names, one of which is voltage follower, which quite accurately describes its operation. Build the circuit as shown in the breadboard layout of Figure 9.10, to see for yourself that this is indeed what happens. All you need do to confirm the results is to measure the input and output voltages at a few different settings, making sure they are the same.

It may seem a bit odd that we have built a circuit, however simple, that appears to do the same job that an ordinary length of

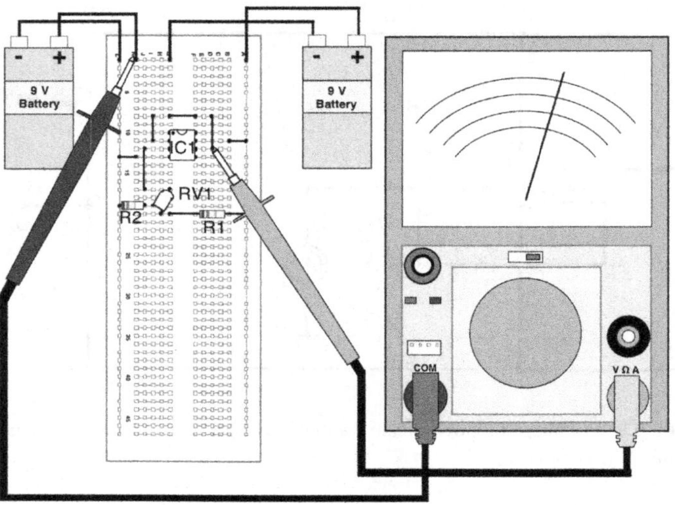

FIGURE 9.10 The breadboard layout of the circuit in Figure 9.9.

wire could perform – that is, the output voltage is the same as the input voltage. But the situation is not quite as simple as that. The circuit we have constructed, although having a gain of only 1, has a very high input resistance. This means that very little current is drawn from the preceding circuit (in this case the preset resistor/voltage divider chain). On the other hand, the circuit has a very low output resistance, which means that it is capable of supplying a substantial current to any following circuits. This property gives rise to another of the circuit's common names – resistance converter.

Hint

In practice the voltage follower is often used when the output of a circuit that is only capable of producing a small, weak current is required to be amplified. The voltage follower is used in the input stage of the amplifier to buffer the current before amplification takes place, thus preventing loading of the preceding circuit. In fact, the term buffer is just another name for the voltage follower.

OFFSET NULLING

Earlier on, it was mentioned that there are two connections to the 741 that are called offset null connections. In most op-amp

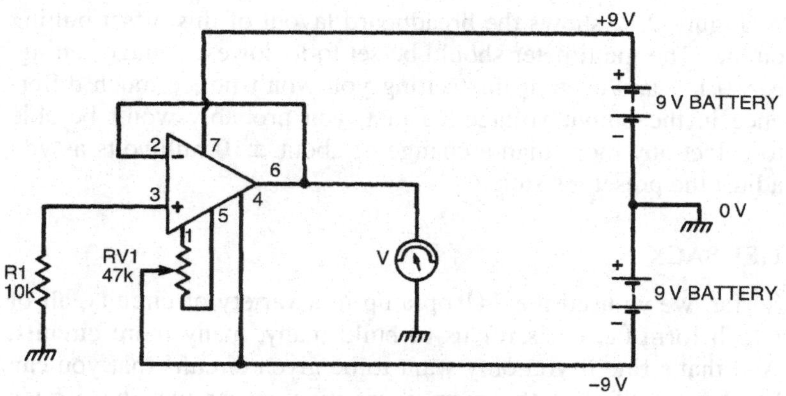

FIGURE 9.11 An experimental circuit to demonstrate the process of offset nulling.

applications these would not be used, but in certain instances, where the level of the output voltage is of critical importance, they are.

When the op-amp has no input voltage applied to it – or, more correctly speaking, the input voltage is 0 V – the output voltage should be the same, i.e. 0 V. Under ideal conditions of manufacture this would be so, but as the 741 is a mass-produced device there are inevitable differences in circuit operation and so the output is rarely exactly 0 V. Temperature differences can also create changes in this output voltage level. The difference in the output from 0 V is known as the offset voltage and is usually of the order of just a few millivolts.

Hint

In the majority of applications this level of offset voltage is no problem and so no action is taken to eliminate it. But the offset null terminals of the 741 may be used to control the level of offset voltage to reduce it to zero.

Figure 9.11 shows a circuit that you can build to see the effects of the process of offset nulling, where a potentiometer has been connected between the two offset null terminals with its wiper connected to the negative supply. Adjusting the wiper controls the offset voltage. You'll see that the circuit is basically a buffer amplifier, like that of Figure 9.9, in which the input voltage to the non-inverting terminal is 0 V.

Figure 9.12 shows the breadboard layout of this offset nulling circuit. The multimeter should be set to its lowest voltage setting, e.g. 0.1 V, and even at this setting you won't notice much difference in the output voltage. In fact, you probably won't be able to detect any more than a change of about ±10 millivolts as you adjust the preset resistor.

GET BACK

So far, we've used the 741 op-amp in a variety of circuits, all of which form the basis for us to build many, many more circuits. And that's fine if you only want to be given circuits that you can build yourself. But the more inquisitive reader may have been wondering how the device can do all of this: after all, it's only through knowledge that greater things can be achieved – to design your own circuits using op-amps, you need to know what makes them tick.

An op-amp can be considered to be a general-purpose amplifier with an extremely high gain. By itself (that is, with no components connected to it) it will have a gain of many thousands. Typical figures for the 741, for example, give a gain of 200,000 at zero frequency, although this varies considerably from device to

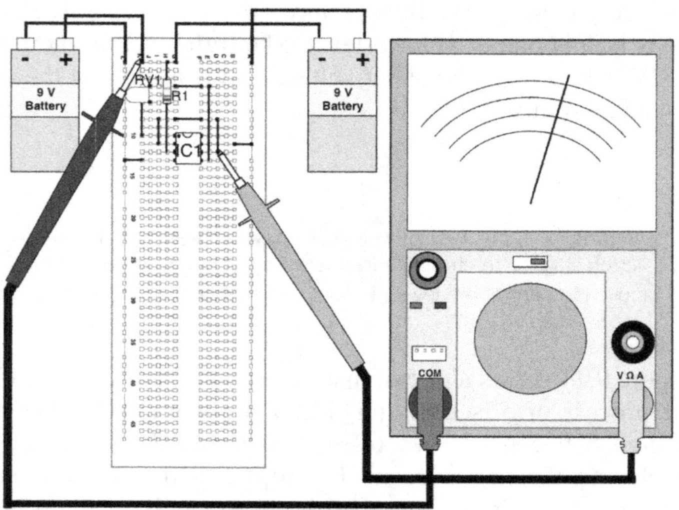

FIGURE 9.12 The breadboard layout of the circuit in Figure 9.11.

device. Now, nobody's suggesting that this sort of gain is particularly useful in itself – can you think of an application in which a gain like this is required? – and besides, as each device has a different gain it would be well nigh impossible to build two circuits with identical properties, let alone the mass-produced thousands of radios, TVs, record players, and so on that use amplifiers. So we need some way of taming this high gain, at the same time as defining its value precisely, so that useful and accurate amplifiers may be designed.

The process used in this taming of op-amps is known as feedback, i.e. part of the output signal from the op-amp is fed back to the input. Look closely again at the circuits we have built so far in this chapter and you'll see that in all cases there is some connection or other from the output back to the input. These connections form the necessary feedback paths that reduce the amplifier's gain to a determined, precise level.

QUIZ

Answers at the end of the book.

1. A typical operational amplifier such as a 741 has:
   - **a.** Two inputs and one output
   - **b.** Two outputs and one input
   - **c.** Three inputs
   - **d.** One input and one output
   - **e.** c and d
   - **f.** None of these.
2. A voltage follower is:
   - **a.** An op-amp circuit with a gain of 100 but which is non-inverting
   - **b.** A unity gain inverting op-amp amplifier
   - **c.** An inverting amplifier, built with an op-amp
   - **d.** b and c
   - **e.** All of these
   - **f.** None of these.
3. An integrated circuit has:
   - **a.** Many transistors
   - **b.** An internal chip of silicon
   - **c.** Pins that allow connection to and from it
   - **d.** All of these
   - **e.** None of these.

4. In the non-inverting amplifier of Figure 9.5, the gain of the circuit when resistor R1 is 100k is:
   a. 1.1
   b. 0.9
   c. 12
   d. 21
   e. None of these.
5. The gain of an op-amp inverting amplifier can be 0.1: true or false?
6. The gain of an op-amp non-inverting amplifier can be 0.1: true or false?

Chapter 10

Digital Integrated Circuits I

If you take a close look at an electronic process – any electronic process – you will always find some part of it that is automatic. By this, I mean that some mechanism is established that controls the process to some extent without human involvement.

Electronic control can be based on either of the two electronics principles we've already mentioned: analog circuits or digital circuits. In the last chapter we saw how analog circuits can be built into integrated circuits. This chapter concentrates on the use of integrated circuits that contain digital circuits – which we call digital integrated circuits.

As we've already seen, integrated circuits (whether analog or digital) are comprised of transistors – lots of 'em. So it's to the transistor that we first turn to in this chapter's look at digital integrated circuits.

When we first considered transistors, we saw that they can operate in only one of two modes: analog (sometimes mistakenly called linear, where the transistor operates over a restricted portion of its characteristic curve) and digital (where the transistor merely acts as an electronic switch – which can be either on or off).

It should be no surprise to learn that analog ICs contain transistors operating in analog mode, while digital ICs contain transistors operating in switching mode. Just occasionally, ICs contain both analog and digital circuits, but these are usually labeled digital anyway.

Figure 10.1 shows a transistor switch and illustrates the general operating principles. It's a single device, with an input at point A and a resultant output at point B. How does it work? Well, we need to mull over a few terms first before we can show this.

LOGICALLY SPEAKING

In the electronic sense, digital signifies that a device has fixed states. Its input is either on or off; its output is also either on or

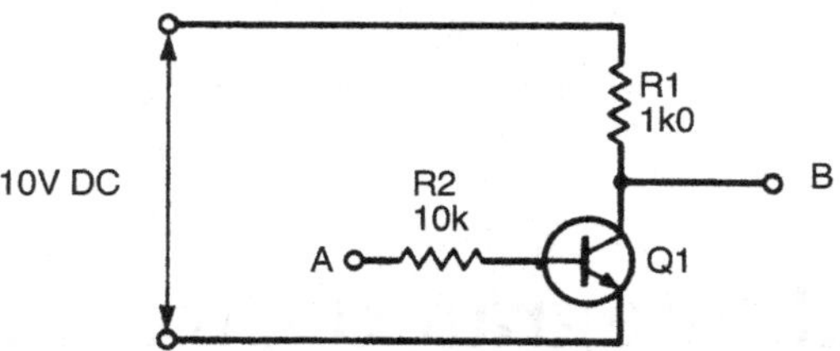

FIGURE 10.1 A simple transistor used as a switch, to form a simple digital circuit.

off. As such there are only really two states, which incidentally gives rise to another term that is often used to describe such electronic circuits – binary digital. Also, it's common to know the states by more simple names, such as on and off, or sometimes logic 0 and logic 1.

Hint

States in an electronic digital circuit are generally indicated by different voltages, with a particular voltage implying one state and another voltage implying the other state. The actual voltages are irrelevant, as long as it's known which voltage implies which state, although by convention typical voltages might be 0V when the state is off or logic 0, and 5 V when the state is on or logic 1. As long as there is a defined voltage for a defined state, it doesn't matter. Further, and as a general rule, just the numbers 0 and 1 can be used to distinguish between the two logic states – whatever the voltages used.

In terms of electronic logic, the circuit of Figure 10.1 is quite simple in operation, and we can work out what it does by considering the transistor's operation. We know from the experiment on page 134 that the presence of a small base current initiates a large collector current through the transistor. We can use this knowledge to calculate what happens in the circuit of Figure 10.1.

If the input at point A is logic 0 (we shall assume that logic 0 is represented by the voltage 0V, and that logic 1 is represented by the voltage 10V) then no base current flows, so no collector current flows – the transistor is switched off. This lack of collector current makes the transistor the equivalent of an extremely high resistance. An equivalent circuit representing this condition is shown in Figure 10.2, where the transistor resistance R_{tr} is 20 M (a typical value for a transistor in the off mode).

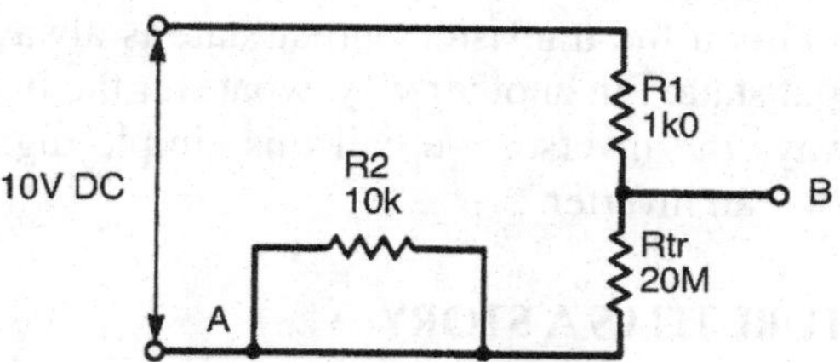

FIGURE 10.2 Equivalent diagram of a single transistor switch, with the transistor base (point A in Figure 10.1) connected to logic 0.

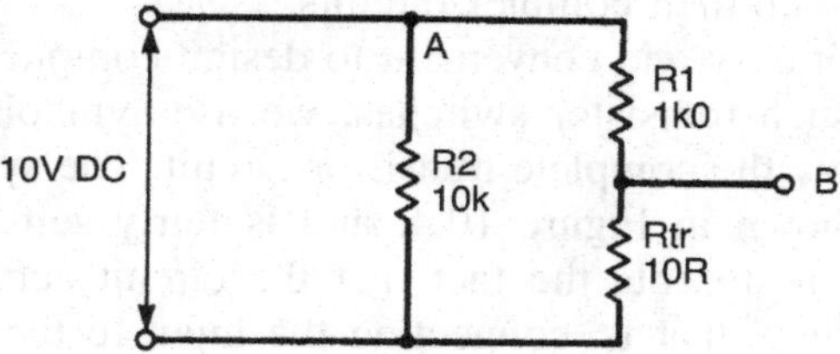

FIGURE 10.3 Equivalent diagram of a single transistor switch, with point A connected to logic 1.

Resistors Rtr and R1 effectively form a voltage divider, the output voltage of which follows the formula:

$$V_B = \frac{R_{tr}}{R_{tr} + R1} \times V_A$$

$$= \frac{20\,M}{20\,M + 1\,k} \times 10$$

which is close enough to being 10V (that is logic 1) to make no difference.

The effect is that output B is connected to logic 1 – in other words, the output at point B is the opposite of that at point A.

If we now consider the transistor with a base current (that is, we connect point A to logic 1, at 10V), the transistor turns on and passes a large collector current, with the effect that it becomes a low resistance. The equivalent circuit is shown in Figure 10.3.

The output voltage at point B is now:

$$V_B = \frac{10}{10 + 1k} \times 10$$

which is as close to 0V (logic 0) as we'll ever get.

So in this circuit the transistor output state is always the opposite of the input state. Put another way, whatever the input state, the output is always the inverse. We call this simple digital circuit – you guessed it – an inverter.

EVERY PICTURE TELLS A STORY

Interestingly, digital circuits of all types (and we'll cover a few over the next few pages) are made up using just this simple and basic configuration. Even whole computers use this basic inverter as the heart of all their complex circuits.

To make it easy and convenient to design complex digital circuits using such transistor switches, we use symbols instead of having to draw the complete transistor circuit. The symbol for an inverter is shown in Figure 10.4, and is fairly self-explanatory. The triangle represents the fact that the circuit acts as what is called a buffer – that is, connecting the input to the output of a preceding stage does not affect or dampen the preceding stage in any way. The dot on the output tells us that an inverting action has taken place, and so the output state is the opposite of the input state in terms of logic level.

To aid our understanding of digital logic circuits, input and output states are often drawn up in a table – known as a truth table – which maps the circuit's output states as its input states change. The truth table for an inverter is shown in Figure 10.5. As you'll see, when its input state is 1 its output state is 0. When its input is 0, its output is 1 – exactly as we've already hypothesized.

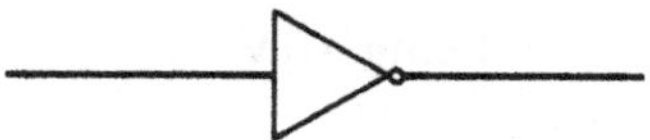

FIGURE 10.4 The symbol of an inverter – the simplest form of digital logic gate.

IN	OUT
1	0
0	1

FIGURE 10.5 Truth table of an inverter – the circuit of Figure 10.1 and the symbol of Figure 10.4.

It's important to remember that, in both a logic symbol and in a truth table, we no longer need to know or care about voltages. All that concerns us is logic states – 0 or 1.

We can build up a simple circuit on a breadboard to investigate the inverter and prove what we've just seen in theory. Figure 10.6 shows the circuit, while Figure 10.7 shows the circuit's breadboard layout. The digital integrated circuit used in Figures 10.6 and 10.7 is actually a collection of six inverters, all accessible via the integrated circuit's pin connections – such an integrated circuit is commonly called a hex inverter, to signify this. We've simply used one of the integrated inverters in our circuit.

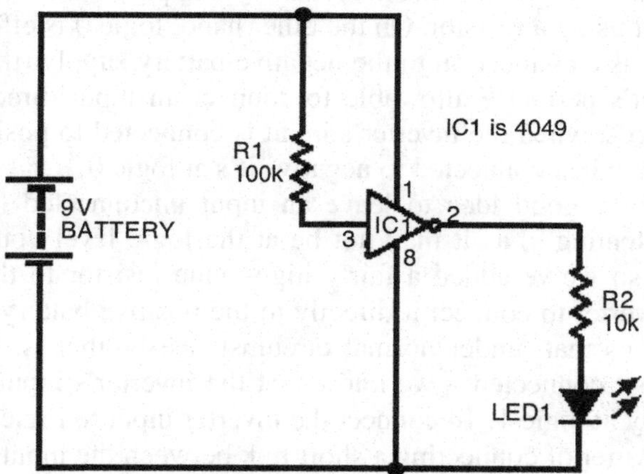

FIGURE 10.6 Circuit of an experiment to discover how an inverter works.

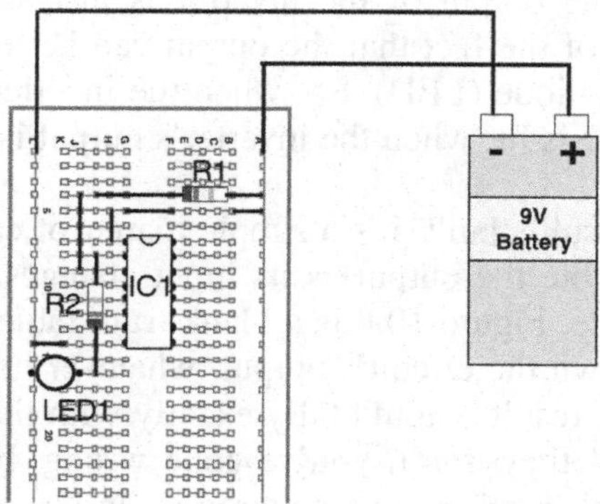

FIGURE 10.7 Possible breadboard layout of the circuit in Figure 10.6.

The integrated circuit used is a type of integrated circuit called a CMOS device, and it's in what's known as the 4000 series of integrated circuits – it's actually known as a 4049. Don't worry about this for now, as we'll look at more devices in the series (and another series, for that matter) later in the chapter. All you need to know for now is that it's a standard dual-in-line (DIL) device, so you must be careful that you position it correctly in the breadboard.

Making sure that the actual integrated inverter used in our circuit has the correct input (that is, either logic 1 or logic 0) is easy. Logic 1 is effectively the same as a connection to the positive battery supply – in practice, though, you should never connect an input "directly" to the positive battery supply; instead, always connect it using a resistor. On the other hand, logic 0 is effectively the same as a connection to the negative battery supply (this time, though, it's perfectly allowable to connect an input directly). In other words, when the inverter's input is connected to positive it's at logic 1, when connected to negative it's at logic 0.

It's not a good idea to leave an input unconnected (what is called "floating"), as it may not be at the logic level you expect it to be, so we've added a fairly high-value resistor to the input of the inverter to connect it directly to the positive battery supply. This means that, under normal circumstances – that is, with no other input connected – we know that the inverter's input is held constantly at logic 1. To connect the inverter input to logic 0, it's a simple matter of connecting a short link between the input and the negative battery supply. This isn't shown in the breadboard layout, but all you need to do is add it when you are ready to perform the experiment.

Measuring the output of the inverter is just as easy, and we take advantage of the fact that the output can be used to light up a light-emitting diode (LED). So, when the inverter's output is at logic 1, the LED is lit; when the inverter's output is at logic 0, the LED is unlit.

Once the circuit's built, it's a simple matter of carrying out the experiment to note the output as its input changes, and completing its truth table. Figure 10.8 is a blank truth table for you to fill in. Just note down the circuit's output, whatever its input state is. Of course, your results should tally exactly with the truth table in Figure 10.5 – if they don't, you've gone wrong, because as I've said before in this book, I can't be wrong, can I?

IN	OUT

FIGURE 10.8 Empty truth table for your results of the experiment in Figure 10.6.

OTHER LOGIC CIRCUITS

The inverter is the first example of a type of digital circuit that's normally called a gate. Other logic gates are, in fact, based on the inverter and are made up from adaptations of it. There's a handful of logic gates that make up all digital circuits, from simple switches through to complex computers.

Other logic gates we need to know about are covered over the next few pages. They are all based on the principles of digital logic we've already looked at. However, before we discuss them, it's important for us to see that they're all based on a particular form of logic, which is essentially mathematical, known as Boolean algebra.

BOOLEAN ALGEBRA

Inputs to and outputs from digital circuits are denoted by letters of the alphabet A, B, C, and so on. Each input or output, as we have seen, can be in one of two states: 0 or 1. So, for example, we can say $A = 1$, $B = 0$, $C = 1$, and so on.

Now, of course, in any digital system, the output will be dependent on the input (or inputs). Let's take as an example the simplest logic gate – the inverter – we've already seen in operation. Figure 10.9 shows it again, along with its truth table, only this time the truth table uses standard Boolean algebra letters A and B. If $A = 1$, then we can see that $B = 0$. Conversely, if $A =$ logic 0, then $B = 1$.

In Boolean algebra we write this fact much more succinctly as:

$$A = \overline{B}$$

where the bar above the letter B indicates what is called a NOT function – that is, the logic state of A is NOT the logic state

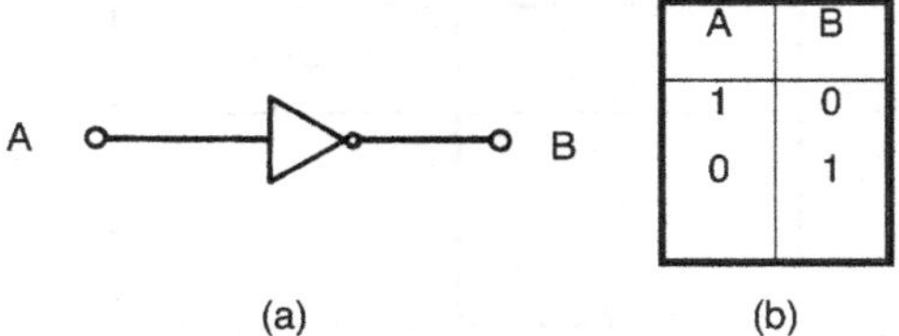

A	B
1	0
0	1

FIGURE 10.9 (a) The symbol for an inverter or NOT gate. (b) Its truth table.

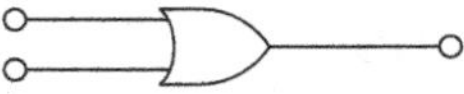

FIGURE 10.10 The circuit symbol for a two-input OR gate.

of B. Alternatively, we could equally well say that B is NOT A, that is:

$$B = \bar{A}$$

Both Boolean statements are true with respect to an inverter, and are actually synonymous. Incidentally, inverters are often called NOT gates for this very reason.

You should be able to see that the Boolean statements for a gate simply allow us a convenient method of writing down the way a gate operates – without the necessity of drawing a gate symbol or forming the gate's truth table. No matter how complex a gate circuit is, if we reduce it to a Boolean statement, it becomes simple to understand.

OTHER LOGIC GATES

The other logic gates we need to know about are every bit as simple as the NOT gate or the inverter. They all have a symbol, they all have a truth table, but most importantly, they all have a Boolean statement we can associate with them.

OR Gate

The symbol of an OR gate is shown in Figure 10.10. To see how an OR gate works, however, we'll conduct an experiment to complete a truth table. The circuit of the experiment is shown in Figure 10.11, while the breadboard layout is in Figure 10.12.

Build up the circuit, and fill in the blank truth table given in Figure 10.13 according to the results of the experiment as you change the circuit's inputs. You should see that the output of the circuit is logic 1 whenever one or other of the circuit's inputs

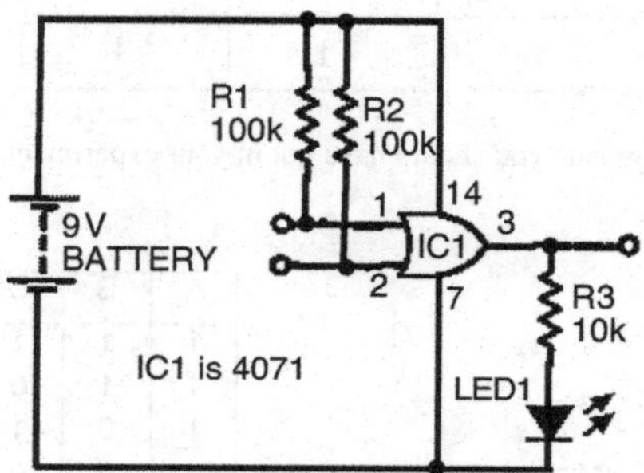

FIGURE 10.11 Experimental circuit to find out how an OR gate works.

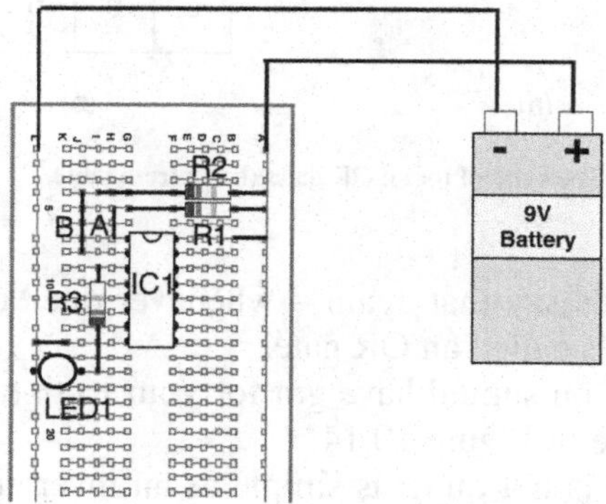

FIGURE 10.12 Possible breadboard layout for the experimental circuit showing OR gate operation in Figure 10.11.

A	B	OUT
0	0	
0	1	
1	0	
1	1	

FIGURE 10.13 A truth table to put the results of your experiment into.

A	B	OUT
0	0	0
0	1	1
1	0	1
1	1	1

FIGURE 10.14 The results you should have got in your experiment.

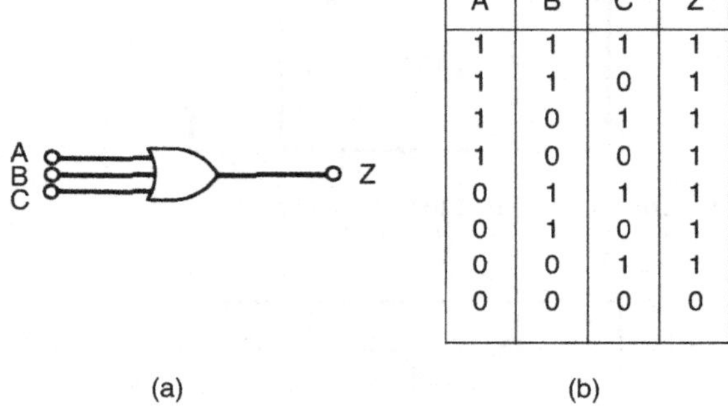

A	B	C	Z
1	1	1	1
1	1	0	1
1	0	1	1
1	0	0	1
0	1	1	1
0	1	0	1
0	0	1	1
0	0	0	0

FIGURE 10.15 (a) The symbol for an OR gate. (b) Its truth table.

are logic 1. Let's say that again – whenever one "or" the other – which is why it's called an OR gate.

The results you should have got for your truth table are shown in the truth table of Figure 10.14.

An OR gate is not quite as simple as an inverter – an inverter, after all, has only one input. An OR gate has more than one input – in theory it can have any number of inputs but, obviously, as the number of inputs increases, the resultant truth table becomes more complicated.

Figure 10.15 shows a three-input OR gate symbol and truth table. Being a three-input gate, its truth table has eight variations. We can calculate this because, being binary digital, the circuit can have $2^3 = 8$ variations of inputs.

The truth table of the OR gate shows that output Z of the gate is 1 when either A OR B OR C is 1. In Boolean terms, this is represented by the statement:

$$Z = A + B + C$$

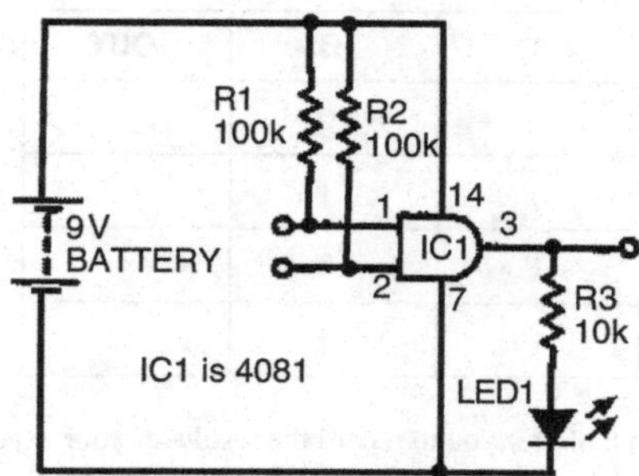

FIGURE 10.16 Circuit to investigate AND gate Boolean operation.

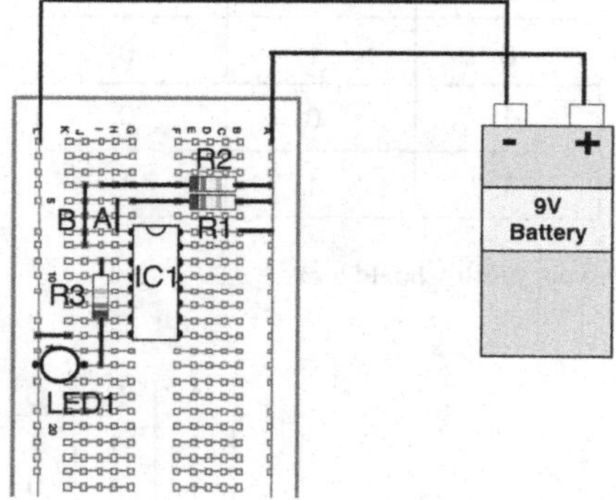

FIGURE 10.17 Possible breadboard layout to investigate the circuit of Figure 10.16.

where "+" indicates the OR function and has nothing whatsoever to do with the more usual arithmetical "plus" sign. Don't be confused: just remember that every time you see the "+" sign in Boolean algebra it means OR.

AND Gate

As you might expect, an AND gate follows the Boolean AND statement. Figure 10.16 shows an AND gate circuit, while its breadboard layout is shown in Figure 10.17, and Figure 10.18 is a blank truth table ready for you to record the experiment's results. As you might already expect, an AND gate is so called because its output is at logic 1 when input 1 AND input 2 are at logic 1. As a result, your experimental truth table should look like the completed truth table shown in Figure 10.19.

A	B	OUT
0	0	
0	1	
1	0	
1	1	

FIGURE 10.18 Truth table for you to record the results of your experiment.

A	B	OUT
0	0	0
0	1	0
1	0	0
1	1	1

FIGURE 10.19 How your results should look.

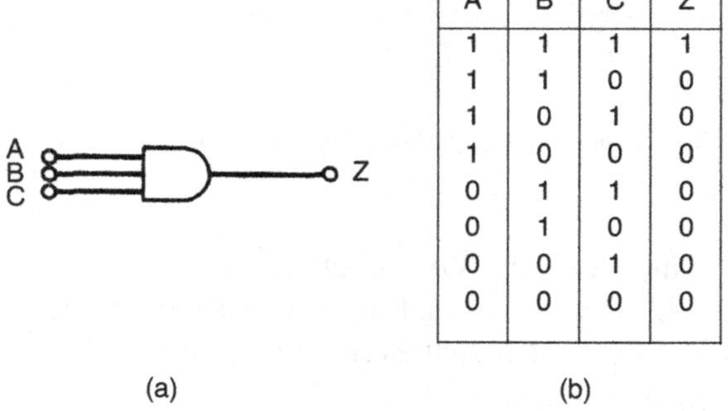

A	B	C	Z
1	1	1	1
1	1	0	0
1	0	1	0
1	0	0	0
0	1	1	0
0	1	0	0
0	0	1	0
0	0	0	0

FIGURE 10.20 (a) A three-input AND gate. (b) Its truth table.

Translating this into a Boolean statement for a three-input AND gate with inputs A, B, and C (shown as a symbol and as a truth table in Figure 10.20), we can see that its output Z is 1, only when inputs A AND B AND C are 1. The Boolean statement is therefore:

$$Z = A \bullet B \bullet C$$

where "•" indicates the Boolean AND expression.

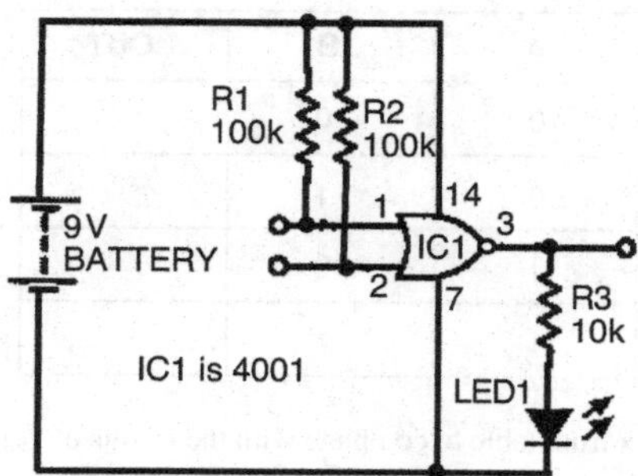

FIGURE 10.21 Circuit to find out the function of a NOR gate.

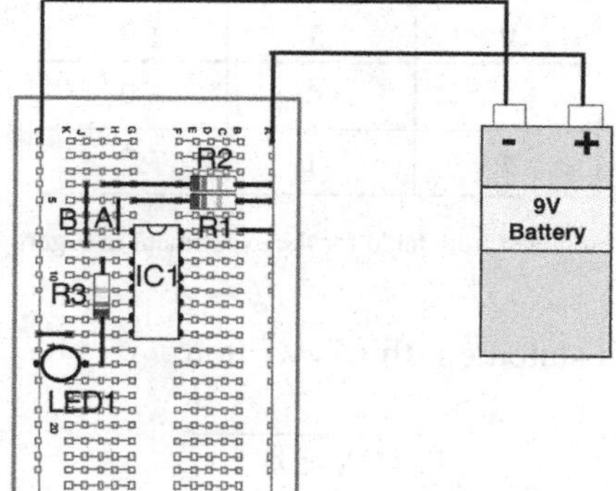

FIGURE 10.22 Possible breadboard layout to build the circuit of Figure 10.21.

NOR Gate

The NOR gate's function can be calculated from its name, "NOR". Effectively, we can think of this as meaning "NOT OR". In other words, it's an OR gate and a NOT gate (that is, an inverter) in series.

The circuit of our experiment to look at NOR gates is shown in Figure 10.21, while a possible breadboard layout for the circuit is given in Figure 10.22. A blank truth table for you to fill in is shown in Figure 10.23. Your completed table should look like the one shown in Figure 10.24.

In Boolean terms, the output of a NOR gate is the exact inverse of the OR gate's output. When applied to the three-input NOR gate and its truth table shown in Figure 10.25, we can see that the NOR gate's output is 1 when neither A NOR B NOR C is 1 – in other words, the exact inverse of the OR gate's output.

A	B	OUT
0	0	
0	1	
1	0	
1	1	

FIGURE 10.23 Blank truth table to complete with the results of your experiment.

A	B	OUT
0	0	1
0	1	0
1	0	0
1	1	0

FIGURE 10.24 A completed truth table for the experiment in Figure 10.21.

As a Boolean statement this is written:

$$Z = \overline{A + B + C}$$

NAND Gate

You should be able to work out for yourself what the NAND in NAND gate stands for: NOT AND. So, if you've followed the last bit about NOR gates, you might also be able to work out its Boolean statement. Nevertheless, it's best to see things for ourselves, and the circuit shown in Figure 10.26 is the experiment we need to carry out to understand what's happening. Figure 10.27 is a possible breadboard layout for the circuit in Figure 10.26. Meanwhile, Figure 10.28 is a blank truth table for you to fill in with your results. Figure 10.29 is the completed truth table for a NAND gate, and yours should be the same (if not, why not, eh?). And, for the sake of completeness, Figure 10.30 shows a three-input NAND gate and its truth table.

Actually, it's quite difficult to express a NAND gate's operation in words, but here goes: it's a gate whose output Z is 1 when NOT A AND NOT B AND NOT C is 1. Err … yeah, right!

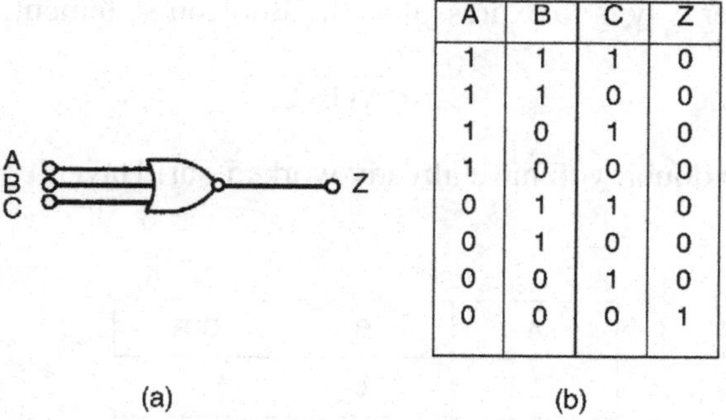

A	B	C	Z
1	1	1	0
1	1	0	0
1	0	1	0
1	0	0	0
0	1	1	0
0	1	0	0
0	0	1	0
0	0	0	1

FIGURE 10.25 (a) A three-input NOR gate. (b) Its truth table.

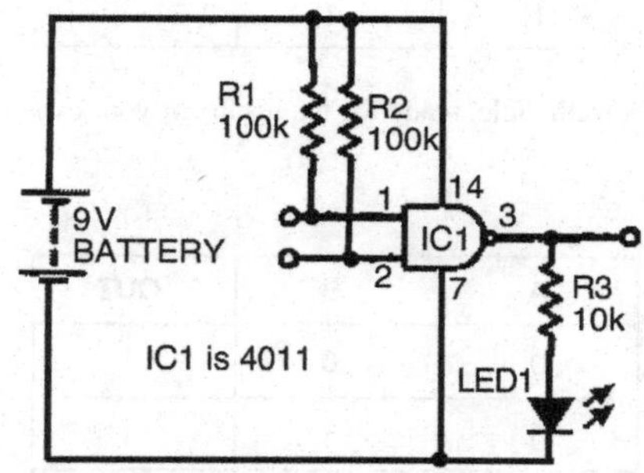

FIGURE 10.26 Circuit to investigate a NAND gate.

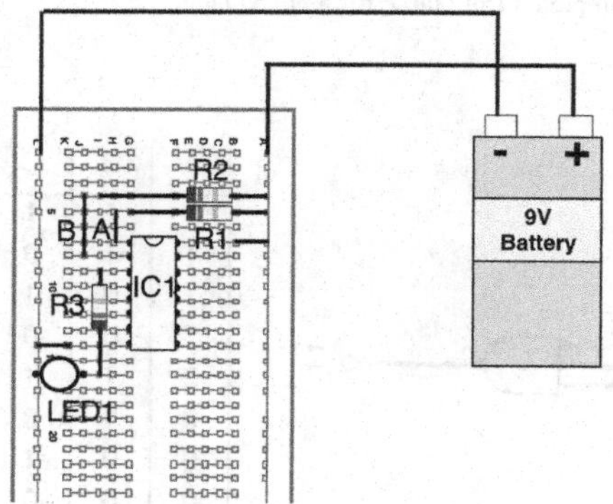

FIGURE 10.27 Breadboard layout to build the circuit of Figure 10.26.

A better way is to express it as the Boolean statement:

$$Z = \overline{A \bullet B \bullet C}$$

which, no doubt, you have already worked out! Haven't you?

A	B	OUT
0	0	
0	1	
1	0	
1	1	

FIGURE 10.28 Blank truth table, ready for the results of your experiment.

A	B	OUT
0	0	1
0	1	1
1	0	1
1	1	0

FIGURE 10.29 Completed truth table for a NAND gate.

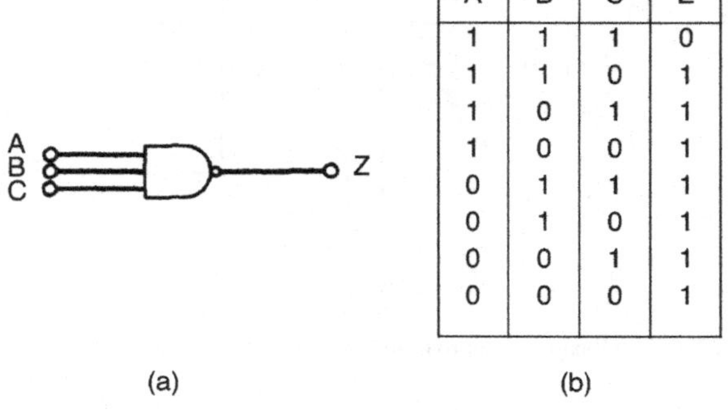

A	B	C	Z
1	1	1	0
1	1	0	1
1	0	1	1
1	0	0	1
0	1	1	1
0	1	0	1
0	0	1	1
0	0	0	1

FIGURE 10.30 A three-input NAND gate. (a) Symbol. (b) Truth table.

SIMPLE, EH?

Earlier on I mentioned that all digital circuits can be made up from the simplest of digital circuits – the inverter. I'm going to prove that now, theoretically at first, then we can do some experiments that show the fact practically.

By expanding the inverter circuit we first looked at back in Figure 10.1 and the following circuits, we can easily create other circuits. Figure 10.31, for example, is merely a transistor operating in the same way that the inverter transistor circuit of Figure 10.1 works. It is, effectively, an inverter with three inputs to its base, rather than just one.

If you look at it carefully, you should be able to figure out how it works. When all of the three inputs A, B, and C are connected to logic 0, the transistor is off, hence the transistor is of very high resistance, which means that the output is at logic 1. However, when any one of the three inputs is connected to logic 1 the transistor turns on, hence becoming low resistance, and the output becomes logic 0. The truth table of Figure 10.32 shows this, and if

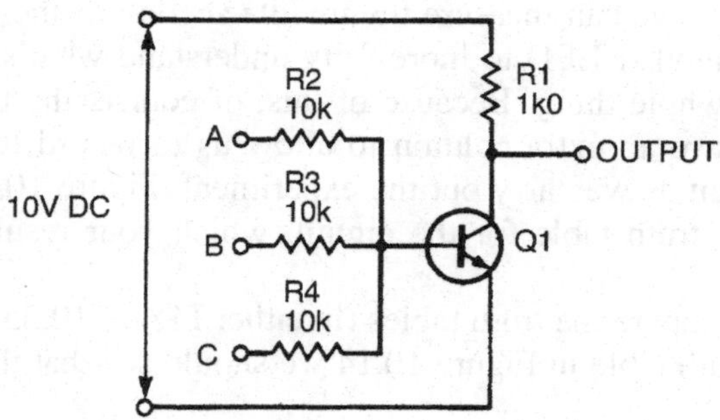

FIGURE 10.31 A transistor switch, with three inputs A, B, and C.

A	B	C	OUTPUT
1	1	1	0
1	1	0	0
1	0	1	0
1	0	0	0
0	1	1	0
0	1	0	0
0	0	1	0
0	0	0	1

FIGURE 10.32 Truth table for the transistor switch circuit of Figure 10.31.

you compare this to the truth table in Figure 10.25 you'll see they are the same. Yes, the three-input transistor switch of Figure 10.31 (which is merely an adapted inverter) is a NOR gate!

THAT OL' BLACK MAGIC!

To show that the inverter is the heart of all other logic gates, we have to consider how other gates are made. We've just seen that we can make a NOR gate from an inverter, but what about the other logic gates?

NOR and NOT Gates Combined

Figure 10.33 is a circuit that combines two logic gates we've already experimented on. They're linked so that the output of one (a NOR gate) can be used as the input of the next (an inverter or NOT gate) to produce a single device.

A possible breadboard layout is shown in Figure 10.34, while an incomplete truth table is shown in Figure 10.35. Note that, as there are two stages to the circuit – that is, two separate logic gates – we can measure the middle section of the circuit by including another LED to more fully understand what's going on inside the whole thing. Because of this, of course, the truth table needs to have an extra column to allow us to record the actions of the circuit as we carry out the experiment. Figure 10.36 shows a complete truth table for the circuit, which your results should mirror.

If we compare the truth tables (in either Figure 10.35 or 10.36) with the truth table in Figure 10.14 we should see that the outputs

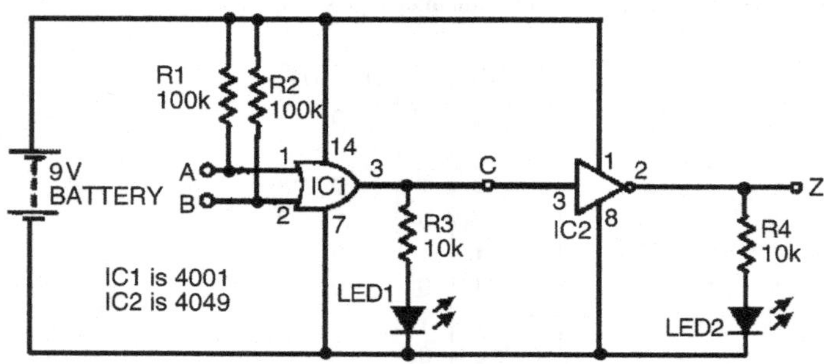

FIGURE 10.33 A simple circuit to create one type of gate (OR) from two other types (NOR and NOT).

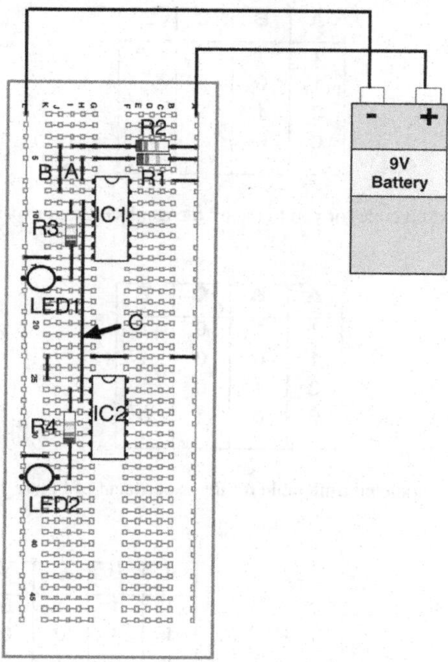

FIGURE 10.34 A breadboard layout for the experiment shown in Figure 10.33.

are the same for the same inputs (OK, yes, the inputs and output are presented in a different order – but the results are identical). In other words, we have used an inverter along with a NOR gate (by merely inverting the output of the NOR gate) to create an OR gate.

For completeness, Figure 10.37 shows a three-input device based on this circuit, together with its truth table.

OR and NOT Gates Combined

In the same vein as the last circuit, Figure 10.38 shows a circuit combining different types of logic gates – this time an OR gate whose inputs have been inverted by NOT gates. The outputs of both NOT gates (hence, also, the inputs of the OR gate) are measured with two LEDs, while the OR gate output is measured with the third LED.

Figure 10.39 shows a breadboard for the circuit for you to follow. Figure 10.40 shows an incomplete truth table for you to

A	B	C	Z
1	1		
1	0		
0	1		
0	0		

FIGURE 10.35 A truth table for you to record the results of your experiment.

A	B	C	Z
1	1	0	1
1	0	0	1
0	1	0	1
0	0	1	0

FIGURE 10.36 A completed truth table for the experiment of Figure 10.33.

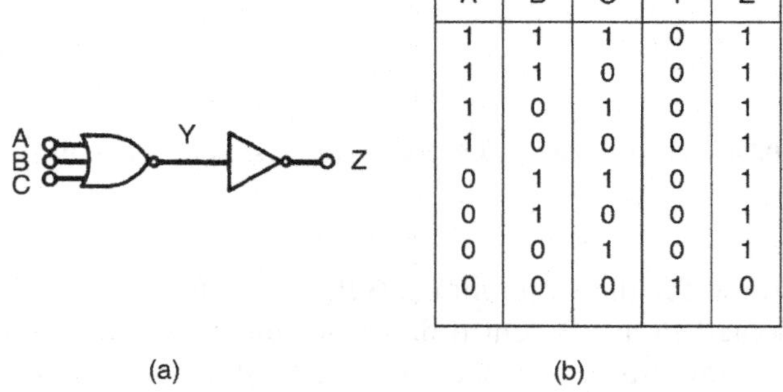

A	B	C	Y	Z
1	1	1	0	1
1	1	0	0	1
1	0	1	0	1
1	0	0	0	1
0	1	1	0	1
0	1	0	0	1
0	0	1	0	1
0	0	0	1	0

FIGURE 10.37 (a) Using an inverter and a NOR gate to create an OR gate. (b) Its truth table to prove this.

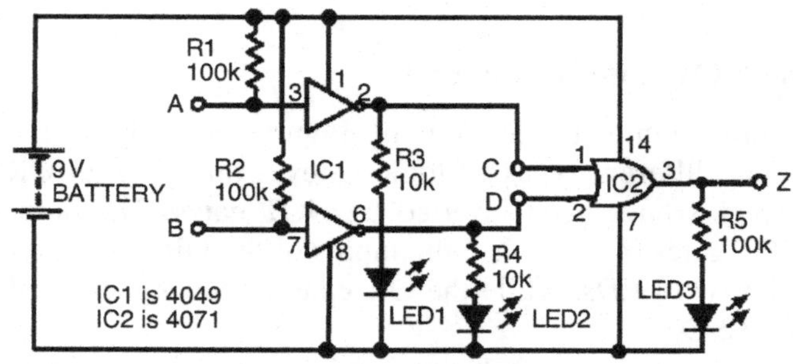

FIGURE 10.38 A circuit to investigate OR and NOT gates.

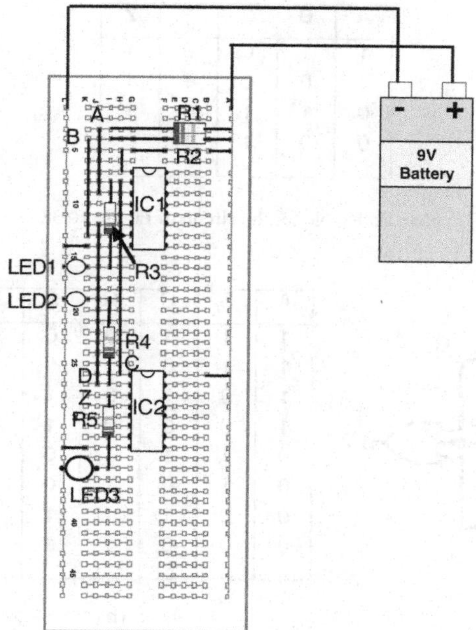

FIGURE 10.39 A possible breadboard layout for the circuit of Figure 10.38.

A	B	C	Y	Z
1	1			
1	0			
0	1			
0	0			

FIGURE 10.40 An incomplete truth table to record your experimental results.

record the results of the experiment, while Figure 10.41 is a complete truth table, which – if everything goes as it should – your results will be identical to.

If you've already been thinking about this, you'll no doubt already have been flicking back through this book's pages to locate which truth table Figures 10.40 and 10.41 are identical to. If not, or if you simply haven't found it, it's the truth table in Figure 10.29 – which is that of a NAND gate.

For the sake of completeness, Figure 10.42(a) shows a three-input variant of this, and the accompanying truth table in Figure 10.42(b) shows the various logic levels in the circuit.

A	B	C	Y	Z
1	1	0	0	1
1	0	0	1	1
0	1	1	0	1
0	0	1	1	0

FIGURE 10.41 Complete truth table for the circuit of Figure 10.38.

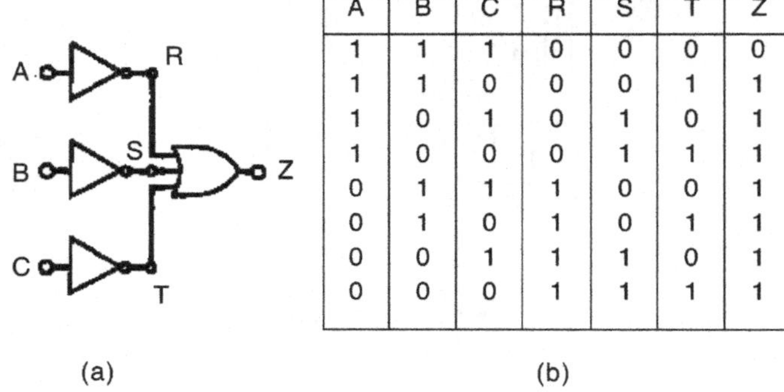

A	B	C	R	S	T	Z
1	1	1	0	0	0	0
1	1	0	0	0	1	1
1	0	1	0	1	0	1
1	0	0	0	1	1	1
0	1	1	1	0	0	1
0	1	0	1	0	1	1
0	0	1	1	1	0	1
0	0	0	1	1	1	1

FIGURE 10.42 (a) A NAND gate, made from OR and NOT gates. (b) Its truth table to prove this.

THE BOOLEAN WAY – IT'S LOGICAL!

Now we already know that the Boolean statement for a NAND gate is:

$$Z = \overline{A \bullet B \bullet C}$$

But we can also derive another Boolean statement from the above circuit, by the very fact that the inputs to the OR gate in Figure 10.42 are inverted before they reach the OR gate, such that the final output Z is equal to logic 1 when NOT A OR NOT B OR NOT C are logic 1.

So, the Boolean statement:

$$Z = \bar{A} + \bar{B} + \bar{C}$$

also expresses the NAND gate.

In other words:

$$\bar{A} + \bar{B} + \bar{C} = \overline{A \bullet B \bullet C}$$

Interesting!

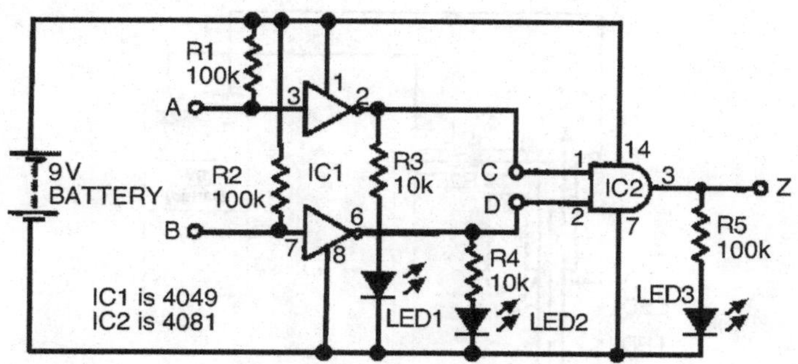

FIGURE 10.43 The experimental circuit to prove that a NOR gate can be created from an AND gate whose inputs are inverted first.

NOR from AND and NOT

And, just for the sake of completion, we can also work out that a NOR gate can be created by inverting the inputs of an AND gate.

As we've done so far in this book, we'll do this first experimentally, building up the circuit and checking results. Then we'll do it mathematically.

Figure 10.43 shows a circuit we can use to do this. It's basically a two-input AND gate, the inputs of which have been inverted by NOT gates. Figure 10.44 shows a possible breadboard layout you can follow to build the experimental circuit on, while Figure 10.45 is an uncompleted truth table for you to complete with your results, and Figure 10.46 is the complete truth table whose results should match yours.

And, for the sake of completeness, Figure 10.47 shows a three-input AND gate with NOT gates at its inputs, and the circuit's truth table, to show that we can make a NOR gate of any number of inputs from an AND gate of that number of inputs together with the requisite number of inverters.

On the other hand, of course, and even better than this, by the same means of deduction we used for other circuits combining logic gates, we can calculate that the standard NOR Boolean statement:

$$Z = \overline{A + B + C}$$

is also the same as another Boolean statement:

$$\overline{A} \bullet \overline{B} \bullet \overline{C}$$

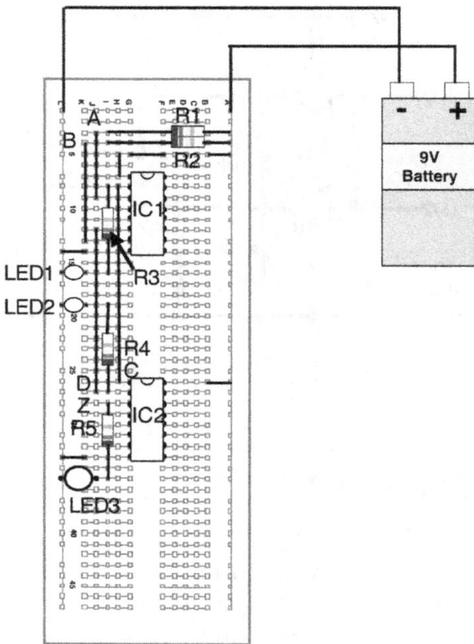

FIGURE 10.44 A breadboard layout for the circuit in Figure 10.43.

A	B	C	Y	Z
1	1			
1	0			
0	1			
0	0			

FIGURE 10.45 An incomplete truth table for you to record your results.

A	B	C	Y	Z
1	1	0	0	0
1	0	0	1	0
0	1	1	0	0
0	0	1	1	1

FIGURE 10.46 A completed truth table for the circuit in Figure 10.43 – showing how we have made a NOR gate from an AND gate with NOT gates at its inputs.

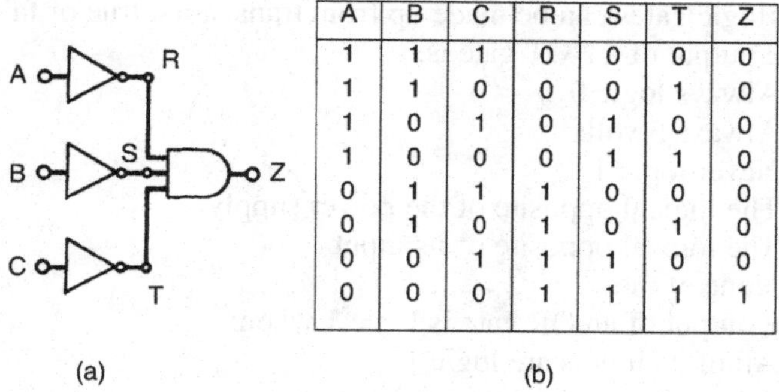

A	B	C	R	S	T	Z
1	1	1	0	0	0	0
1	1	0	0	0	1	0
1	0	1	0	1	0	0
1	0	0	0	1	1	0
0	1	1	1	0	0	0
0	1	0	1	0	1	0
0	0	1	1	1	0	0
0	0	0	1	1	1	1

FIGURE 10.47 (a) A three-input AND gate converted to a NOR gate with NOT gates. (b) The circuit's truth table.

In other words:

$$\overline{A} \bullet \overline{B} \bullet \overline{C} = \overline{A + B + C}$$

And – there – we have just proved mathematically that you can make a NOR gate from AND and NOT gates.

So, there's no doubt about it, all digital electronic logic gates can be created from a number of other digital electronic logic gates.

What's more, it's not beyond the realms of possibility to work out for ourselves that these very logic gates, that can make any other logic gates, can be combined, and combined again, to make up ever more complex circuits. And, indeed they are – as we'll see in the next chapter! For now, have a go at the quiz that follows, to see if you've been taking notice … which I'm sure you have. Just prove it!

QUIZ

Answers at the end of the book.

1. In the circuit symbol for a NOT gate:
  a. The input is applied to the side with a circle
  b. The output comes from the side with a circle
  c. The circle implies that the signal is turned upside down
  d. The triangle represents the three signals that must be applied
  e. b and c
  f. All of these
  g. None of these.

2. All logic gates can be made up from transistors: true or false?
3. The output of a NOT gate is:
   a. Always logic 0
   b. Always 5 volts
   c. Never logic 1
   d. The logical opposite of the power supply
   e. The logical opposite of its input
   f. None of these.
4. The output of an OR gate is logic 1 when:
   a. All of its inputs are logic 1
   b. Any of its inputs are logic 1
   c. None of its inputs are logic 1
   d. a and b.
5. The output of a NOR gate is logic 1 when:
   a. All of its inputs are logic 1
   b. 10 milliamps
   c. Any of its inputs are logic 1
   d. This is a trick question, the output is always logic 1
   e. None of these.
6. When the output of a gate is used as the input of a NOT gate, the final output is:
   a. Always the inverse of the output of the first gate
   b. Always the same as the output of the first gate
   c. Can only be calculated mathematically by binary logic
   d. All of these
   e. None of these
   f. In the lap of the gods.
7. If the inputs to an OR gate are inverted with NOT gates, the overall gate is that of:
   a. A NOT gate
   b. A NOR gate
   c. A NAND gate
   d. a and b
   e. b and c
   f. None of these.
8. Any logic gate can be created from other logic gates: true or false?

Chapter 11

Digital Integrated Circuits II

In the last chapter we saw how electronic logic gates can be constructed from other electronic logic gates; however, as far as the hobbyist or the person starting out in electronics is concerned, knowing that logic gates can be created this way is one thing. No one in their right mind would actually go about creating anything other than the simplest of logic gates like that, because it's just too time-consuming and too expensive. ICs represent the only sensible option.

Now we're going to take a look at some IC logic gates, in the form of the same sort of ICs we've already studied and experimented with. The ICs shown over the next few pages represent only the tip of the digital IC iceberg though, as I've only shown some of the most common and simplest. The digital circuits we'll look at later in this chapter are all available in one IC form or another as well, so the list of available ICs is very long indeed – not just the few shown here.

IC SERIES

For most people, there are two families of ICs that are used in digital electronics. The first is a very easy to handle variety (in other words, they can't be easily damaged). On the other hand, they require a stable operating voltage of 5V, which isn't all that easy to generate without a power supply.

The second family of digital ICs will operate over a range of voltages (typically, a 9V battery does the job exceedingly well), but are a little more easily damaged. Nevertheless, handled carefully and correctly, both families work well.

7400 Series

The 7400 series of digital ICs is the variety that requires a fixed 5V power supply. While there are many, many devices in the

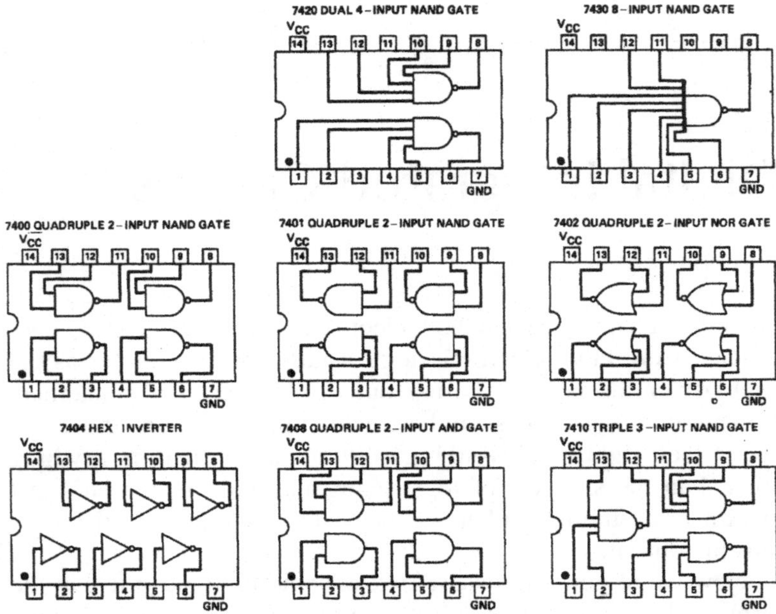

FIGURE 11.1 Some integrated circuits from the 74 series of transistor–transistor logic (TTL) digital devices.

family only a few are of interest here. Before I explain about the ICs in detail though, a brief word about the 7400 series numbering.

The first two digits of any device in the family (that is, 74–) indicate that the device is a member of the family. The 7400 series is a family of digital ICs known as transistor–transistor logic (TTL) devices. The last two (or sometimes three) digits (−00) indicate which IC this is in the family. In other words, a 7400 (which happens to be an IC with four two-input NAND gates) is different to a 7402 (which has four two-input NOR gates).

A selection of ICs in the 7400 series is shown in Figure 11.1. Note that those shown – as I've already said – represent only a small fraction of the complete family.

4000 Series

Like the 7400 series, the devices in the 4000 series are represented by their numbers. Note, though, that the numbers between the two families do not correspond – in other words, a 4001 does

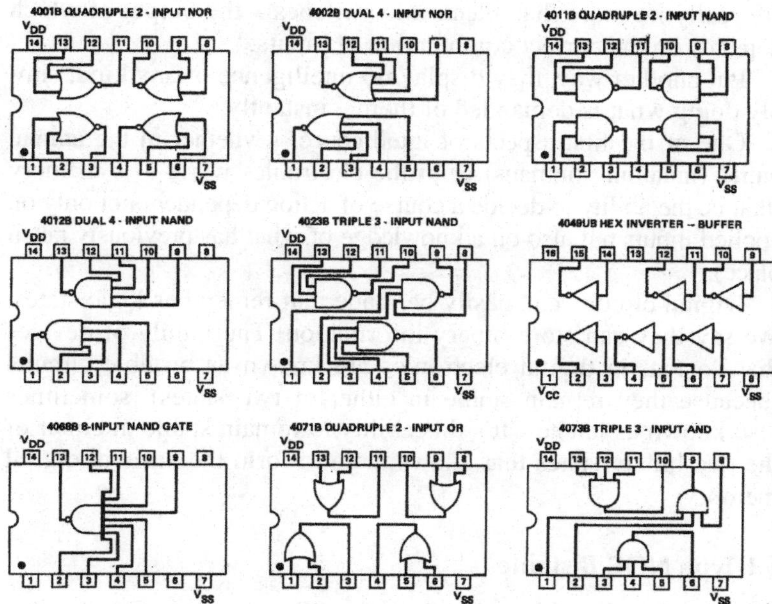

FIGURE 11.2 Some common devices from the 4000 series of digital integrated circuits – note that these cannot be directly interchanged with devices from the 74 series of TTL digital integrated circuits.

not contain the same logic gates as a 7401. A small part of the 4000 series is shown in Figure 11.2.

Once we have a family (well, two families actually) of digital integrated circuits that contain logic gates, it's fairly obvious that building more complex digital circuits becomes easier. It is merely a matter of using the gates within these integrated circuits to build more complex circuits. On the other hand, further devices in both ranges of integrated circuits are themselves more complex, containing circuits built from the gates themselves.

SHUTTING THE STABLE GATE

All the digital electronic circuits we've looked at so far function more or less instantly. If certain inputs are present, a circuit will produce an output that is defined by the Boolean statement for the circuit – it doesn't matter how complex the circuit is, a combination of inputs will produce a certain output.

For this reason, such circuits are known as combinational. While they may be used in quite complex digital circuits they are really no more than electronic switches – the output of which depends on the correct combination of inputs.

Put another way, they display no intelligence of any kind, simply doing what is demanded of them – instantly.

One of the first aspects of intelligence – whether in the animal world (including humans) or in the electronics world – is memory (that is, the ability to decide a course of action dependent not only on applied inputs but also on a knowledge of what has previously taken place).

Digital circuits can easily be made that remember logic states. We say they can store binary information. The family of devices that do that in digital electronics are known as bistable circuits (because they remain stable in either of two states), sometimes also known as latches. It's this ability to remain stable in either of the two binary states that allows them to form the basis of digital memory.

SR-Type NOR Bistable

The simplest bistable or latch is the SR-type (sometimes called RS-type) bistable, the operation of which acts like two gates connected together, as in Figure 11.3. The circuit is constructed so that the two gates are cross-coupled – the output of the first is connected to the input of the second, and the output of the second is connected to the input of the first.

There are two inputs to the circuit and also two outputs. The inputs are labeled S and R (hence, the reason why the circuits are called SR-type), which stand for "Set" and "Reset". Now, it so happens that one output is, in most instances, the inverse of the other, so for convenience one output (by convention) is labeled Q and the other output (also by convention) is labeled $\overline{Q}$ (called bar-Q) to show this.

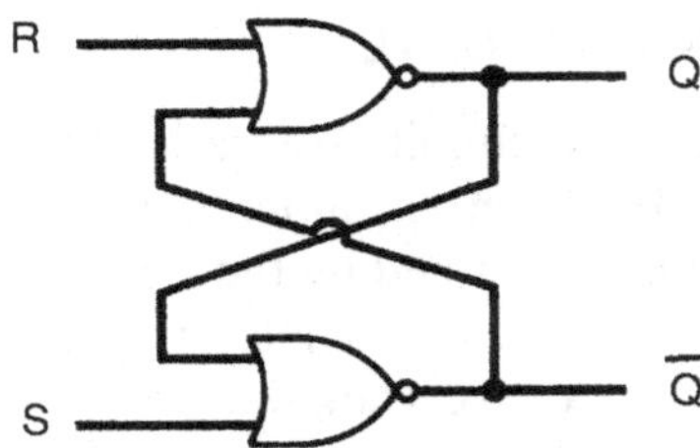

FIGURE 11.3 An SR-type bistable circuit, comprising two cross-coupled NOR gates.

Both inputs to the SR-type NOR gate bistable should normally be at logic 0. As one input (either S or R) changes to logic 1, the output of that gate goes to logic 0. As the gates are cross-coupled, this is fed back to the second input of the other gate, which makes the output of the other gate go to logic 1. This, in turn, is fed back to the second input of the first gate, forcing its output to remain at logic 0 – even (and this is the important point) after the original input is removed. This is an important point!

Applying another logic 1 input to the first gate has no further effect on the circuit – it remains stable, or latched, in this state.

On the other hand, a logic 1 input applied to the other gate's input causes the same set of circumstances to occur but in the other direction, causing the circuit to become stable, or latch, in the other way.

This sort of switching backwards and forwards in two alternate but stable states accounts for yet another term that is often applied to bistable circuits – they are commonly called "flip-flops".

The SR-type NOR bistable operation can be summarized in a truth table, in a similar way that non-bistable or combinational logic gates can be summarized. However, because we are trying to tabulate inputs and outputs that change with time, then strictly speaking we should call this a "function table", and such a function table for the SR-type NOR bistable is shown in Figure 11.4.

Account has been taken in the function table for changing from one logic state to the other, by the symbol [_|‾], which indicates that the input state changes from a logic 0 to logic 1. Also, we no longer write the logic states as 1 and 0, but H (meaning high) and L (meaning low).

The outputs indicate what happens as the input states change on these occasions. You can see that if input S goes high (that

S	R	Q	$\overline{Q}$
_|‾	L	H	L
L	_|‾	H	H
L	L	Q_0	$\overline{Q}_0$
H	H	L	L

FIGURE 11.4 The function table of the SR-type NOR bistable circuit shown in Figure 11.3.

is, from 0 to 1) while input R is low, then the output Q will be high. If, however, input S then returns low, the output will be Q_0 – which simply means that it remains at what it was. If input R then goes high while S remains low, the output changes to low and remains in that state even if R goes low again.

The last line of the function table is shaded to indicate that this condition is best avoided – simply because both outputs are the same. Designers of digital circuits would normally take steps to ensure that this condition did not occur in their logic circuits because confusion in the form of uncertain outcomes can obviously arise. In other words (and, yes, I know that it was an awful lot of words), we have constructed a form of electronic memory, one of the outputs of which can be set to a high logic state by application of a positive-going pulse to one of its inputs. It's a small but very significant advance – changing combinational logic gates into bistable circuits that can be used together to create sequences of logic level variations. In effect, what we have now done is to cross over from combinational logic circuits to a new type of logic circuits known as sequential logic circuits, with this simple but highly significant and very important addition of memory. Whatever we want to call them – bistables, latches, or flip-flops – they are sequential, and they can be used to create even more complex logic circuits.

SR-Type NAND Bistable

In Figure 11.3 the gates are both NOR gates, but they can be just as simply NAND gates, as shown in Figure 11.5. This gives us an SR-type bistable made from NAND gates. Easy enough – but there are some important considerations.

The most important consideration is that both inputs of the SR-type NAND bistable should normally be at logic 1 – this is in direct contrast to the SR-type NOR bistable, where the inputs needed to be normally at logic 0. Because of this, the inputs are considered to be inverted in the circuit, shown as such in Figure 11.5.

With both inputs of the circuit of Figure 11.5 at logic 1, the circuit is stable, with both NAND gate outputs at logic 0. When one input goes to logic 1, the output of that NAND gate goes to logic 1. This is applied to the other NAND gate's second input, so the output of the second NAND gate will go to logic 0. This output is in turn fed to the first NAND gate's second input and thus its output is forced to remain at logic 1.

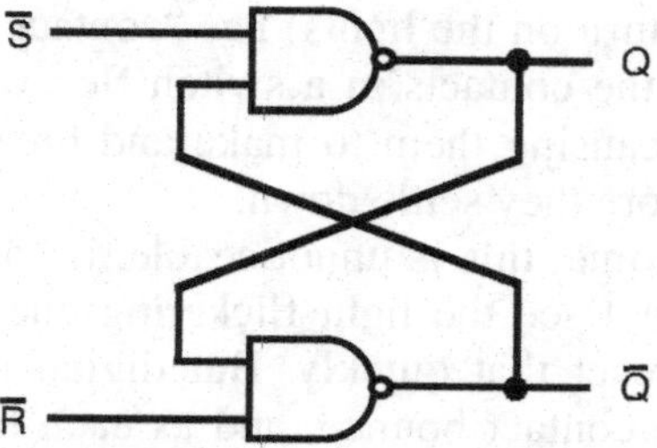

FIGURE 11.5 An SR-type bistable circuit comprising two cross-coupled NAND gates – note the inverted inputs when compared with Figure 11.3.

$\overline{S}$	$\overline{R}$	Q	$\overline{Q}$
‾|_	H	H	L
H	‾|_	L	H
H	H	Q_0	$\overline{Q}_0$
L	L	H	H

FIGURE 11.6 A function table for an SR-type NAND bistable circuit, such as that in Figure 11.5.

Applying a further logic 0 to the first NAND gate has no further effect on the circuit – it remains stable. However, when a logic 0 is applied to the second NAND gate's input this causes the same reaction in the other direction. So the circuit has two stable states, and hence is a bistable. Figure 11.6 shows the function table for this SR-type NAND bistable.

Just as in the SR-type NOR bistable, there is a possibility that both outputs can be forced to be the same level, which produces an indeterminate outcome. In the SR-type NAND bistable, this is when both inputs are at logic 0. So, circuit designers must ensure this situation does not occur in their logic circuits.

While these simple gate circuits are really all that's required to build bistables in theory, practical concerns require that something is added to them before they work properly and as expected. The problem is that the very act of applying changing logic levels to its inputs suffers from the mechanical limitations of the connecting method. Any mechanical switch (even one you might use in your home to turn on the lights) has "contact bounce" – a phenomenon where the contacts in a switch flex while the switchover takes place, causing them to make and break the connection several times before they settle down.

Now, in the home, this is unnoticeable, because it happens so quickly you cannot see the light flickering due to it, even if the light itself can react that quickly. But digital circuits are quick enough to notice contact bounce, and as each bounce is counted as a pulse the circuit could have reacted many, many times to just one flick of the switch. While we were merely changing input states and watching output states of combinational logic circuits in the last chapter, this really didn't matter. All we were wanting was a fixed output for a set of fixed inputs. In sequential circuits like bistables – and the other sequential circuits we'll be looking at in this chapter, as well as others – contact bounce is a real problem, as we want a sequence of inputs. The wrong sequence – given by contact bounces – will give us the wrong results. Simple as that!

The likes of simple bistables like the SR-type NAND bistable give us an ideal and effective solution to the problems created by contact bounce. We've just discussed how once one of the NAND gates in the circuit is switched by application of one logic 0 pulse at either its S or R input, it then doesn't matter if any further pulses are applied to the input. So, a switch that provides the pulse can make as many contact bounces as it likes, the circuit will not be affected further.

Figure 11.7 shows the simple addition of a switch and two resistors to the SR-type NAND bistable, to give a circuit that can be used in other circuits to provide specific logic pulses that have absolutely no contact bounce.

Switch SW1 in the circuit of Figure 11.7 is a simple single-pole, double-throw (SPDT) switch, and should have break-before-make contacts (in other words, when switching from one connection to the other, its switch contacts will disconnect from one connection before it connects to the other – so there is a point in the switch motion when all three contacts are disconnected from each other). The switch can be either a push-button type or a conventional flick-type switch.

Operation is fairly simple. At rest, one of the NAND gate inputs connects through the switch to logic 0, while the other NAND gate's input is connected through a resistor to logic 1. As the switch is operated, the second NAND gate's input is

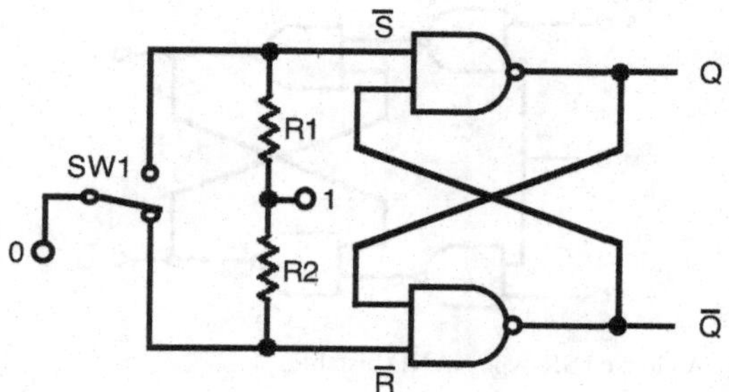

FIGURE 11.7 Simple circuit to prevent contact bounce.

connected to logic 0 though the switch, and the circuit operates as previously described like a standard SR-type NAND bistable. It doesn't matter whether there is contact bounce or not at this time, as repeated pulses do not affect the bistable.

When the switch is operated again, so the first NAND gate's input is connected to logic 0 and the bistable jumps to its other stable state. Again, contact bounce will make no difference to it.

One or both of the outputs of the simple de-bouncing circuit can be applied to the inputs of following circuitry, safe in the knowledge that contact bounce will have no effect. The requirement for no contact bounce is true for any sequential circuit actually. So this very simple solution can be used not just in the simple circuits we see here in this chapter, but for all sequential circuits. Whenever a sequential logic circuit requires a manually pulsed input, this circuit can be used.

OTHER BISTABLES

There is more than one type of bistable – the SR type is in reality only the simplest. They all derive from the SR-type bistable, however, and so it's the SR-type bistable that we'll use as the basic building block when making them.

The Clocked SR-Type Bistable

Figure 11.8 shows the clocked SR-type bistable. It is a simple adaptation of the basic SR-type NAND bistable, in that it features an extra pair of NAND gates, which not only allows inputs that do not require inverting, but also allows a clocked input to be part of the circuit's controlling mechanism.

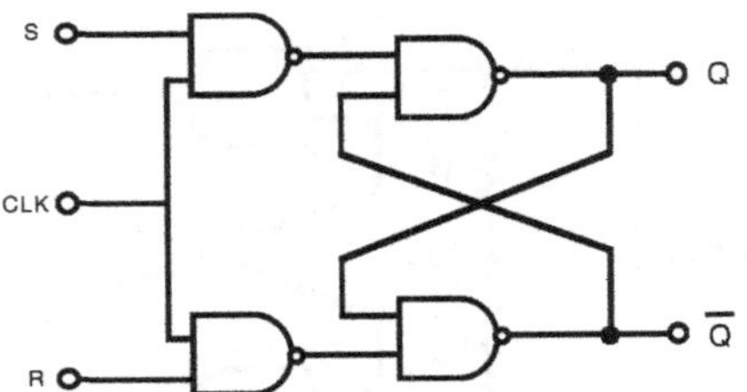

FIGURE 11.8 A clocked SR-type NAND bistable.

Now, a clock circuit can be made from something like a 555 astable multi-vibrator (as we saw in Chapter 5), and how it is made is unimportant to us here. What is important, on the other hand, is that a clock means that several bistables can be synchronized, all in time with the clock.

The clocked SR-type NAND bistable circuit here functions basically as before, except that the outputs can only change states while the clock input CLK is at logic 1. When the clock input CLK is at logic 0, it doesn't matter what logic levels are applied to the S and R inputs, there will be no effect on the bistable.

Thus, with this simple addition, we have created a bistable with a controlling input. Unless the controlling input is at logic 1, nothing else can happen. This can be a very useful thing in electronics where, say, a number of things need to be counted over a period of time. For example, by clocking the circuit for, say, 1 hour, the number of cars passing a checkpoint in that time can be counted electronically.

The D-Type Bistable

A variation on the basic clocked SR-type bistable is the "data" bistable, or D-type bistable. A circuit is shown in Figure 11.9.

Effectively, the output of the first NAND gate (which would be the inverted S input) is used as the input to the second (which would be the R input). The single remaining input is given the label D.

In operation, the Q output of the D-type bistable will always follow the logic level at D while the clock signal CLK is at logic 1. When the clock falls to logic 0, however, the output Q remains at the last state of the D input prior to the clock changing.

This is quite a useful circuit, and is relatively simple in terms of both circuitry and operation. Like the simple SR-type bistable,

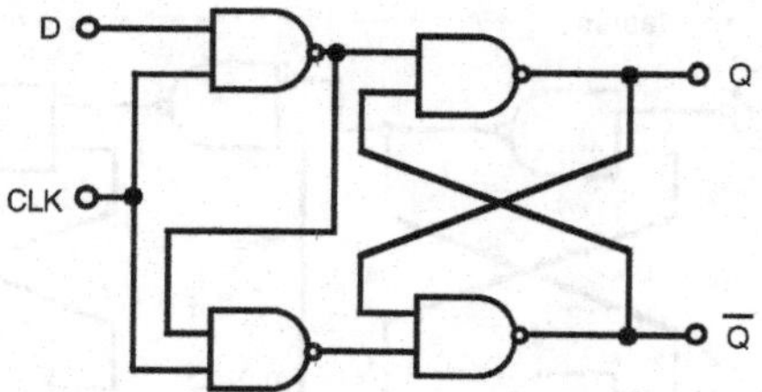

FIGURE 11.9 A D-type bistable.

it can be used to measure input signal variations over a defined period of time.

The Edge-Triggered SR-Type Bistable

However, often it's necessary that things are registered at a precise instant of time, not over a period. So the circuits of Figures 11.8 and 11.9 cannot be used. A moderately simple adaptation to ensure a circuit only registers input changes at an instant is all that's required – it's basically a combination of two identical clocked SR-type NAND bistables, each operating on opposite halves of the clock signal – and the resultant circuit is shown in Figure 11.10.

Such a combination of two bistables is often called a "master–slave" bistable, because the input bistable operates as the master section, while the output bistable is slaved to the master during half of each clock cycle.

An important component is the inverter, which connects the two bistables in the circuit. This ensures that the bistables are enabled during opposite half cycles of the clock signal.

Assuming that the clock input CLK is at logic 0 initially, the S and R inputs cannot yet affect the master bistable's operation. However, when the clock input CLK goes to logic 1 the S and R inputs are now able to control the master bistable in the same way as they do in Figure 11.8. As the inverter has inverted the clock signal, though, the slave bistable's inputs (which are formed by the outputs of the master bistable) now have no effect on the slave's outputs. In short, although the outputs of the master bistable may have changed, they do not yet have any effect on the slave bistable.

When the clock input CLK falls back to logic 0 the master bistable once again is no longer controlled by its S and R inputs. At the same time, however, the inverted clock signal now allows

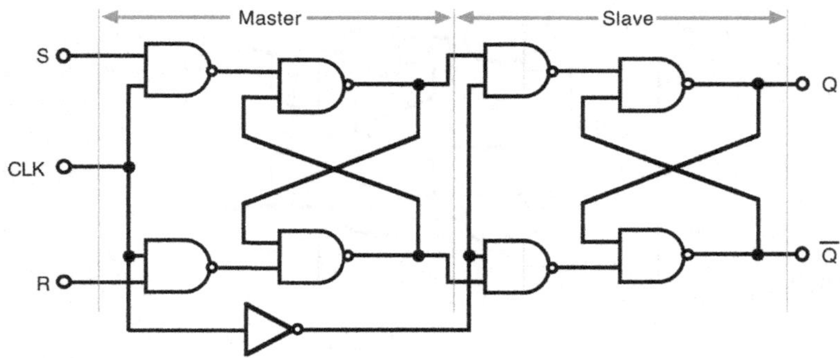

FIGURE 11.10 An edge-triggered SR-type NAND bistable.

the slave bistable's inputs to control the slave bistable. In other words, the final outputs of the circuits can only change state as the clock signal CLK falls from logic 1 to logic 0. This change of state from logic 1 to logic 0 is commonly called the "falling edge", and the overall circuit is generically known as an "edge-triggered" bistable.

This is an extremely important point in electronic terms. By creating this master–slave bistable arrangement to make the bistable edge-triggered, we are able to control precisely when the bistable changes state. As a benefit, this also makes sure that there is plenty of time for the master and slave bistables comprising the overall bistable to respond to the input signals – although things in logic circuits change and respond quickly, they do not happen instantly and still do take a finite time. The master–slave arrangement takes account of and caters for this small but finite time.

The JK-Type Bistable

One other problem that we've already encountered with our basic bistables isn't yet catered for though – the indeterminate output that can occur in a bistable if both S and R inputs are logic 1 at the moment when the clock signal falls from logic 1 to logic 0.

So, to prevent this happening, it's a matter of preventing both S and R inputs from being at logic 1 at the same time as the clock signal falls from logic 1 to logic 0. We do this by adding some feedback from the slave bistable to the master bistable, and creating new inputs (labeled J and K). The circuit of such a JK-type bistable to perform this function is shown in Figure 11.11.

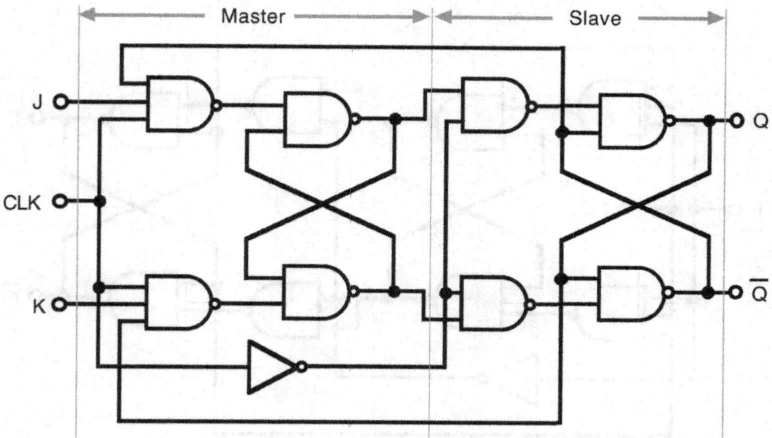

FIGURE 11.11 A JK-type bistable.

As with the edge-triggered master–slave SR-type bistable, the outputs only change on the falling edge of the clock CLK signal, so the inputs (J and K now, not S and R) control the output states at that time. However, the feedback from the final output stage back to the input stage ensures that one of the two inputs is always disabled – so the master bistable cannot change state back and forth while the clock input is at logic 1. Instead, the enabled input can only change the master bistable state once, after which no further change of states can occur.

Because the JK-type bistable is completely predictable in this manner, under all circuit conditions, the JK-type bistable is the preferred minimum bistable device for logic circuit designers. That's not to say that SR-type bistables cannot be used, and in fact they do have their purposes, but the important point is that circuit designers have to be aware of their limitations, ensuring that unpredictable outcomes are not allowed and so are designed out of the circuit.

T-Type Bistable

One particular mode of operation of the JK-type bistable is of importance here. If both J and K inputs are held at logic 1, the outputs of the master bistable will change state with each rising edge of the clock signal and the final outputs will change state for each falling edge. This fact is used as the basis of the T-type bistable

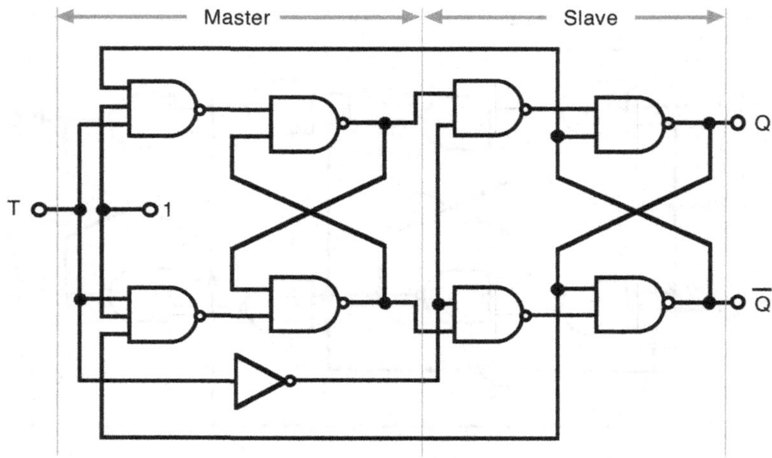

FIGURE 11.12 A T-type bistable.

(where T stands for "toggle") shown in Figure 11.12. The lone T input of this circuit is really just the CLK inputs of other bistables.

And that concludes our look at how logic bistable circuits are made up from basic logic gates. The important point about all of this, though, is that by taking basic combinational logic gates and joining them in what are fairly basic circuits, we have created sequential circuits – that is, logic circuits that have an element of memory!

OPEN THE BLACK BOX

Just as we found earlier that we could create logic gates with collections of transistors, but decided we could draw logic gates with their own symbols (rather than actually drawing all the internal transistors involved), so there are symbols for the various types of bistable – which don't show the actual logic gates (or, indeed, the individual transistors) involved.

Figure 11.13, for example, shows the symbol for a clocked, edge-triggered, SR-type bistable, such as the circuit of Figure 11.10. It's just a box, with the bistable's main inputs shown on the box. Note, though, that it doesn't show whether the SR-type bistable is created from NAND gates or NOR gates (or whatever gates, or indeed whatever transistors, actually – that's all very irrelevant!). All you need to remember is that it's just a box showing inputs and outputs – and that it acts as a clocked, edge-triggered, SR-type bistable.

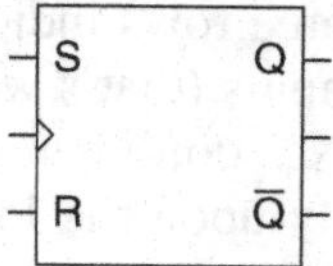

FIGURE 11.13 A clocked, edge-triggered, SR-type bistable symbol.

S	R	Q	$\overline{Q}$
↑	L	H	L
L	↑	H	H
L	L	Q_0	$\overline{Q}_0$
H	H	L	L

FIGURE 11.14 The function table of the clocked, edge-triggered, SR-type bistable of Figure 11.13.

Of note is the symbol used for the clock input, which merely represents the fact that it is edge triggered. Its function table is shown in Figure 11.14.

Such an approach of simplifying circuits is very common in electronics, and we've come across it before – the black-box approach. The approach relies on the fact that we don't need to know how a complex circuit is constructed – we only need to know its function.

Every single one of the other logic gate types we've seen in this chapter can also be represented in black-box block form by symbols and, to prove it, we'll consider them all now. All you have to do is remember that if necessary they can (and sometimes are) built using logic gates, but you don't need to know how to construct them that way because – yes, you've guessed it – they can be bought ready-built in IC form, in the same series (7400 and 4000) we've already used and are familar with.

First is the D-type bistable. Its circuit symbol is shown in Figure 11.15, while its corresponding function table is in Figure 11.16.

Note that the inputs $\overline{PR}$ (preset) and $\overline{CL}$ (clear) are included in this D-type bistable (unlike the D-type bistable shown in Figure 11.9). These perform similar functions to the S and R inputs of the SR-type bistable. You should now be able to understand the function table. The first three rows indicate that the $\overline{PR}$ and $\overline{CL}$ inputs override all other inputs (that's why the clock and D inputs are shown as X: because X = don't care). Thus, the outputs Q and $\overline{Q}$ are dependent primarily upon the logic states of the $\overline{PR}$ and $\overline{CL}$ inputs.

However, when the $\overline{PR}$ and $\overline{CL}$ inputs are both high, control of output becomes the prerogative of the clock and D inputs. For example, when the clock is low, the output Q = Q_0. That is, it remains the same as it was – no matter what the logic state of the D input (as D = X). On the leading edge of the clock pulse, however, output Q follows the logic state of the input D, whether high or low.

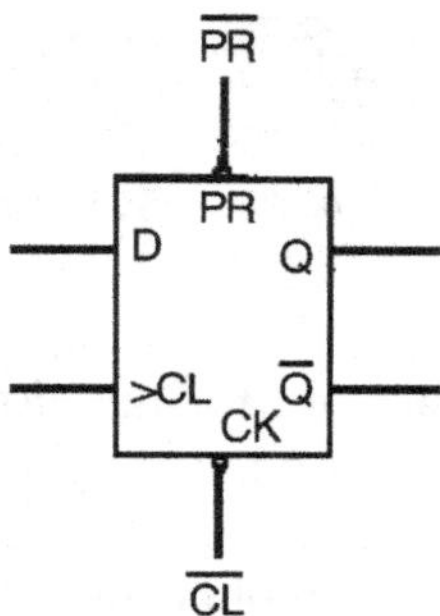

FIGURE 11.15 A D-type bistable symbol.

$\overline{PR}$	$\overline{CL}$	CK	D	Q	$\overline{Q}$
L	H	X	X	H	L
H	L	X	X	L	H
L	L	X	X	H	H
H	H	↑	H	H	L
H	H	↑	L	L	H
H	H	L	X	Q_0	$\overline{Q_0}$

FIGURE 11.16 The function table of a D-type bistable, such as the one shown in Figure 11.15.

JK-Type Bistable Symbol

The JK-type bistable symbol is shown in Figure 11.17, while its function table is shown in Figure 11.18.

There's a couple of points to note about this JK-type bistable symbol:

- Operations take place on the falling edge of the applied clock pulse.
- The last row of the function table shows a bistable operation known as toggling – in which the output state changes from whatever state it is, to the other state. This, you will of course remember, is how the T-type bistable of Figure 11.12 is created.

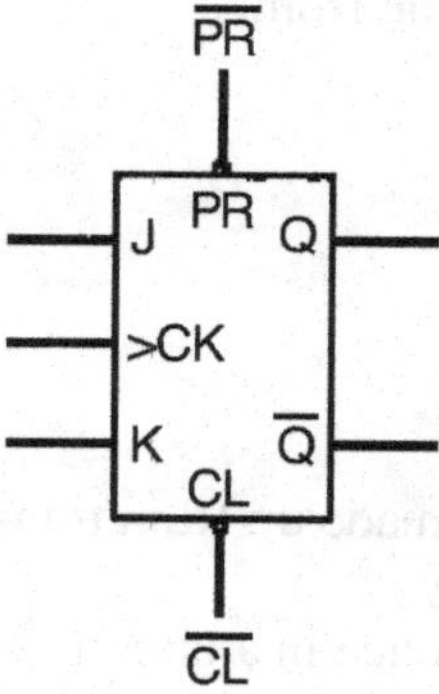

FIGURE 11.17 A JK-type bistable.

$\overline{PR}$	$\overline{CL}$	CK	J	K	Q	$\overline{Q}$
L	H	X	X	X	H	L
H	L	X	X	X	L	H
L	L	X	X	X	H	H
H	H	⎍	L	L	Q_0	$\overline{Q}_0$
H	H	⎍	H	L	H	L
H	H	⎍	L	H	L	H
H	H	⎍	H	H	TOGGLE	

FIGURE 11.18 The function table of a JK master–slave bistable, such as that shown in Figure 11.17.

THE END OF THE DIGITAL LINE …

So there we go. We've taken an in-depth look at digital integrated circuits and the electronic logic that they comprise. They allow a very different perspective on the electronics world than all the components and techniques we've met before, and they open up electronics to areas that couldn't be reached without them.

To check your understanding of digital integrated circuits and the circuits they allow, make sure you do the quiz that follows!

QUIZ

Answers at the end of the book.

1. A bistable can be made from:
   - **a.** Two NOR gates
   - **b.** Two NAND gates
   - **c.** An IC
   - **d.** a and b
   - **e.** b and c
   - **f.** All of these
   - **g.** None of these.
2. All bistables can be made up from transistors: true or false?
3. A function table:
   - **a.** Shows contact bounce in a circuit
   - **b.** Is like a truth table for sequential circuits
   - **c.** Never applies to bistables made from logic gates
   - **d.** Only applies to IC bistables
   - **e.** a and b
   - **f.** None of these.
4. Contact bounce:
   - **a.** Is a myth
   - **b.** Is impossible to eliminate practically
   - **c.** Always occurs on Saturday
   - **d.** Is none of these.
5. All clocked bistables are edge triggered: true or false?
6. One way of making an edge-triggered bistable:
   - **a.** Is to combine two bistables in a master–slave arrangement
   - **b.** Is to adjust the clock signal so that it only has edges
   - **c.** Is to disconnect the battery
   - **d.** All of these
   - **e.** None of these.

7. The T in T-type bistable stands for:
   a. Total
   b. Typical
   c. Transient
   d. Toggle
   e. Two sugars and milk
   f. None of these.

8. The JK-type bistable is useful as:
   a. It needs no power supply
   b. All possible outputs can be predetermined logically
   c. A doorstop
   d. It allows fewer inputs than an SR-type bistable
   e. a and c
   f. b and c
   g. None of these.

Chapter 12

Soldering

The things you need when reading this chapter are generally tools, though there's a few small items of hardware, and you're going to need a few spare components – resistors, say – to use when you're practicing soldering. Parts – although officially these are all tools – you need are:

- Soldering iron
- Soldering iron stand
- Sponge
- Soldering mat (or plasterboard square)
- Solder – read through the chapter to decide if you need lead-free solder, lead-based solder, or both
- Abrasive fiberglass pencil
- Scrubbing block
- Soldering iron bit tip cleaner
- Printed circuit board or stripboard
- Terminal pins
- Connecting lead – a selection of multi-strand and single-strand insulated wire, and uninsulated single-strand wire (for links)
- Desolder braid
- Desolder tool.

Whatever you do in electronics – whatever you make, whatever you build – at some point you will discover that the single most important process in the whole business is soldering. Soldering is – quite literally – the glue that holds all the other processes in electronics together. You can learn about electrons whizzing around a circuit; you can learn about the various components in that circuit, and how they control those whizzing electrons; you can study every book ever written about electronics, you can master and obtain any university degree or qualification the education system can throw at you – BUT without soldering the actual parts of a circuit together, all that knowledge and learning is useless! It is all merely knowledge, learning, and theory. Soldering transforms all that knowledge, learning, and theory into something practical, something physical, something you can get hold of and use – in essence, something that works.

WHAT DOES SOLDERING DO?

Soldering has two basic functions:

- To hold electronics components in their positions on a printed circuit board.
- To connect components to other components – both physically and/or electrically – in the circuit.

OK – So What's a Printed Circuit Board?

I've just mentioned a printed circuit board, but what exactly is a printed circuit board? Well, look inside any modern electronics appliance (television, computer, mobile phone, etc.) or even many electrical appliances (washing machine, iron, kettle, etc.) and you'll see a printed circuit board – often known by the abbreviation PCB.

A printed circuit board is a thin baseboard (about 1.5 mm) of insulating material such as resin-bonded paper or fiberglass, with an even thinner layer of copper (about 0.2 mm) on one or both surfaces. (If copper is only on one surface it's then known as single-sided printed circuit board; if copper is on both surfaces it's known as double-sided printed circuit board.) The copper on the surface of a printed circuit board has been printed as a circuit (yes, OK, that's why it's called printed circuit board – geddit?), so that components on the printed circuit board can be soldered to the copper, and thus be connected to other components similarly soldered. Photo 12.1 shows a fairly modern printed circuit board to show you what they

Take Note

The term printed isn't exactly an accurate description of how the copper on the surface of a printed circuit board is formed. In fact, all printed circuit boards start life with a complete layer of copper on one or both sides of the insulating board. Then, unwanted copper is removed from the board, leaving the wanted copper pattern behind. Typically, this copper removal is usually – though not always – done by etching the copper away using strong chemicals.

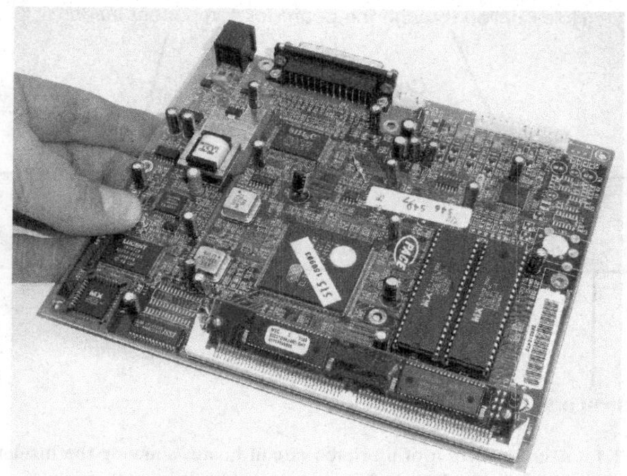

PHOTO 12.1 A printed circuit board – the type you might find in a computer or other modern electronic appliance.

PHOTO 12.2 A basic printed circuit board – here you can see the copper surface that allows connections to be made between components.

look like. The printed circuit board shown is quite a complex one, with hundreds of components – from a computer actually – but the printed circuit board in a washing machine, say, may only hold a handful of components. Photo 12.2 shows how the copper on a printed circuit board comprises a pattern of copper – sometimes called the copper track – rather than a solid layer. This pattern or track is the key to making connections between components.

Figure 12.1 shows a cross-section of a simple printed circuit board. In it you can see the insulating board, the copper track, and the holes for component leads. Components fit to the printed

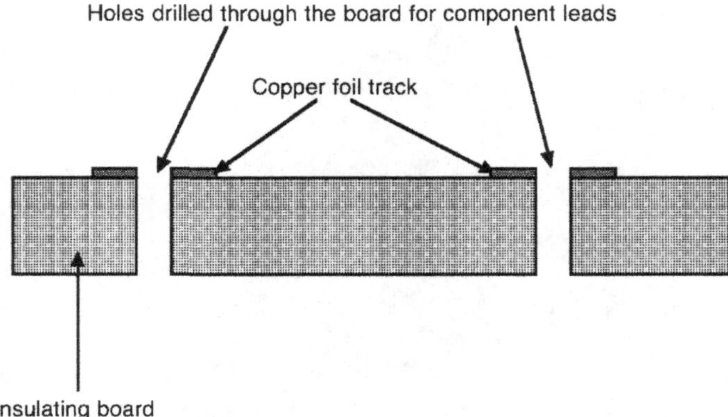

FIGURE 12.1 A cross-section of a printed circuit board, showing the insulating board, the copper track, and holes in the board for component leads to go through.

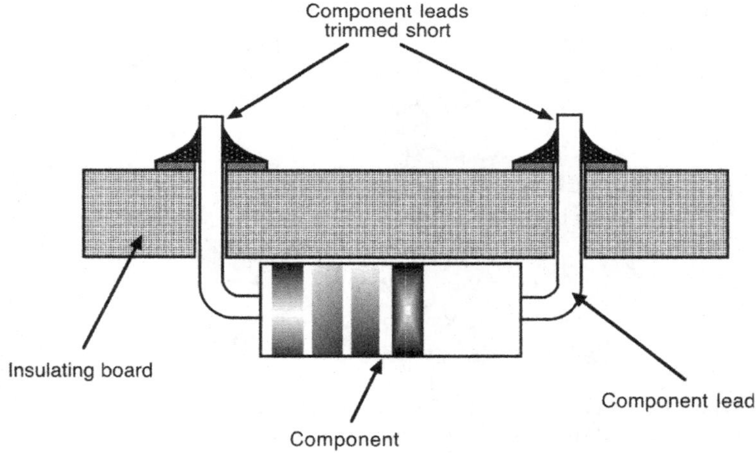

FIGURE 12.2 A cross-section of a printed circuit board, showing a component fitted. The component leads pass through the holes in the board and are soldered to the copper track.

circuit quite easily. Their leads are inserted through the board holes, and are then soldered to the copper track. Figure 12.2 shows how this works. In terms of the amateur enthusiast in electronics, simple (and relatively inexpensive) hand-tools are all that are required in this soldering process – we'll look at these, and how to use them, later.

TOO FAR, TOO FAST

Whoooaaaahhh! Steady there! We've jumped the gun a little here, in an effort to show you how the soldering of components to a printed circuit board works. We now need to step back and consider soldering – and indeed, solder itself – as theoretical and practical principles. We also need to learn how to actually solder ourselves. It's not difficult – but there is a knack to it (some say an art, even) that some people don't always find too easy to learn. That's the aim of this final chapter of the book – to teach us enough about solder and soldering to be able to do it ourselves.

WHAT IS SOLDER?

Solder is a metallic compound that has a low enough melting point – somewhere around 200°C – that it can be heated by small hand-tools (called soldering irons) till it melts and thus joins other metals (in electronics terms: component leads, copper track) together when it cools and re-solidifies.

Solder has been used for thousands of years. The Romans were great users of it. Historically at least, one of the main constituents of solder has been the metal lead. Indeed, the Latin name of lead – plumbum – gives a clue to one of solder's main uses, in plumbing, to join water pipes and fittings.

When solder is used to join two parts (in plumbing or in electronics) together, the assembly is called a soldered joint. In plumbing this occurs between, say, two copper water pipes. In electronics, the term soldered joint is usually used to refer to where a component lead is soldered to the copper track of a printed circuit board. The copper track of a printed circuit board allows a soldered joint made this way to connect to other soldered joints (also formed by other component leads and the copper track) so that the complete electronic circuit can be made up on the printed circuit board. As a by-product, the rigidity of the soldered joints making up the circuit also means that all components are also held securely in place.

What is Solder Made Of?

Except in its very earliest use, when the metal lead itself was used in pure form, solder has always been a mixture of two or more metals. There is a simple reason for this – pure metal lead melts at a temperature of 327°C. An alloy of only 38% lead, together with 62% tin, melts at only 183°C, a great deal lower than the melting point of pure lead. So it's possible to solder lead pipes together using such a solder mixture without actually melting the lead pipes themselves.

This 38% lead/62% tin solder alloy has become known as the eutectic composition, and its melting point of 183°C (the lowest melting point of any tin/lead variations of solder alloy) is therefore known as the eutectic point. It's not hard to understand that – purely in terms of soldering component leads to printed circuit board copper tracks in electronics – this eutectic solder alloy has inevitably been viewed historically as a very useful, if not the ideal, solder alloy.

But that has all changed!

The metal lead, of course, quite rightly, is viewed as a toxic, poisonous, substance. So much so that as of 1 July 2006, European manufacturers or importers of electronic and electrical equipment are no longer allowed to use lead (along with various other similar toxic materials) in the products they make. The regulation that came into force on 1 July 2006 within the European Union is called RoHS (which stands for the restriction on the use of hazardous substances). Products such as solder, or components, that meet the RoHS regulations are often known as RoHS compliant.

Obviously, RoHS makes a difference to the materials used to make the solder we now use in electronics. The term lead-free solder has been coined for the solders now used in industrial manufacturing of electrical and electronic equipment, and many alloys have been researched to provide essentially similar – but not quite identical – properties to old-fashioned lead-based solders.

As far as us, the amateur electronics enthusiasts – i.e. non-commercial makers of electronic equipment – are concerned, lead-free solder is not essential (the RoHS regulations only cover commercial manufacturers or importers), and we can choose to use lead-based solder for now, if we wish. However, lead-based solder is becoming increasingly hard to locate, as most electronics suppliers have long ceased to stock it. That's not to say you can't use it if you can find lead-based solder and, indeed, it has one or two advantages over lead-free solder for us as ordinary users. However, this book, its author and publishers and so on, in respect of the RoHS regulations, cannot condone the use of lead-based solder, and if a reader chooses to use it then that remains their decision.

So What's the Difference?

In terms of general use, there is little obvious difference between lead-based and lead-free solder. Both fulfill the two basic functions of solder we looked at earlier:

- To hold electronics components in their positions on a printed circuit board.
- To connect components to other components – both physically and/or electrically – in the circuit.

Both lead-based and lead-free solders are alloys of metals, which exhibit a lower melting point than the metals they are intended to joint together. Both are also usable with the same (or, at least, very similar) hand-tools. Finally, both also are available in the same formats for use by amateur enthusiasts.

So, to all intents and purposes, lead-based and lead-free solders should be directly interchangeable. So what exactly is the difference?

Well, first, it should be obvious that the alloys used in the two types of solder must be different. Lead-based solder uses tin and lead. Lead-free solder uses various mixtures of tin, silver, bismuth, zinc, copper, and so on as their alloy parts. The elements used in a solder alloy make a real difference to the price of the solder, and unfortunately the best solders for our purposes tend to be the most expensive.

Second, lead-based solder has a lower melting temperature than any lead-free solder. As we already know, eutectic lead-based solder has a melting temperature of 183°C. The equivalent lead-free solder (depending on the constituent metals) has a melting temperature of around 30–40°C higher than this. This may not mean much at first sight, but this extra heat can potentially damage some of the very components you may solder into a printed circuit board. A typical lead-free solder available off the shelf from electronics component suppliers is a tin/silver/copper alloy (95.5% tin/4% silver/0.5% copper), which has a melting temperature of around 217°C. Another lead-free solder commonly available is just a tin/copper alloy (99.3% tin/0.7% copper) and is slightly cheaper than the previous one, but has a melting temperature of about 227°C. As you might expect, the lead-free solder with the lowest melting temperature is more expensive.

Third, lead-free solders – as far as readers of this will be concerned, at least – aren't quite as easy to use as lead-based solder.

This is partly because of the higher melting temperature and the resultant potential component damage, but also because molten lead-free solder simply does not flow as readily as lead-based solder. Flow is the term we use to denote the process of the solder melting and joining to the component and copper track surfaces.

Lead-Based vs. Lead-Free

OK, so now we know the main differences between lead-based and lead-free solder. But when can we use either type, and when should we not use a particular type?

Well, whether you use lead-based or lead-free solder in a new project you undertake is your personal choice (currently, at least – as we know, the RoHS regulations merely specify that manufacturers or importers of electrical or electronic equipment must not use lead-based solder). Lead-based solder is definitely easier to use for the beginner to electronics, and typically gives noticeably better results. It does, however, contain lead, and you may want to consider the potential health consequences of its use.

On the other hand, if you ever service or repair older electrical or electronic equipment, which was built before the RoHS regulations came into effect on 1 July 2006, you may find that it was built using lead-based solder. If that is the case, you simply will not be able to use lead-free solder, as the two types physically do not allow use together. Even in the unlikely event you manage to create a soldered joint between lead-free solder and a joint originally made using lead-based solder, the new, resultant joint will be very much weaker and will probably deteriorate rapidly towards failure. So, if you ever service or repair such equipment, you will need to carry at least a small reel of lead-based solder.

Another point to bear in mind is that the component leads of components that have been available to buy since 1 July 2006 usually come pre-tinned. But, obviously, to be RoHS compliant, the pre-tinning must use lead-free solder as the tinning agent, which means that you will need to use lead-free solder to solder them into a printed circuit board. However, components before that date may have been pre-tinned with lead-based solder, so those of us (particularly old-timers like me!) with stocks of components may have a mixture of components with lead-based and lead-free pre-tinning! Life is not always easy.

THE GOOD SOLDERED JOINT

Whatever the actual solder alloy used, the principles of making a soldered joint are the same. An area of metal is heated, then solder is applied, whereupon the solder melts and wets the metal surface to form a joint. Note that the term wet is used, and in some respects means much the same as a liquid such as water wetting a surface – it flows over the surface.

Wetting in soldering terms, on the other hand – unlike water wetting a surface – is a process where the solder comes into direct metallic contact with the metals to be soldered together into the joint. At the junction between the solder and the metal, a specific compound of both the solder and the metal is formed. This compound – known as the intermetallic compound – occurs in an extremely thin layer between the solder and the metal. Sometimes, the bond at that point is called the intermetallic bond. Interestingly, once this intermetallic compound has been created, it cannot be removed – in other words, you simply cannot remove solder from a metal once it has been wetted!

Take Note

When water wets a surface, you can usually remove the water, leaving the surface as it was before it was wet. When solder wets a surface, the surface is changed permanently – an intermetallic bond is created, which cannot be removed.

Figure 12.3 shows a cross-section of a soldered joint, illustrating the main details. Photo 12.3 shows an actual micro-section photograph of a soldered joint, in which can be seen the solder, the metal – copper – and the intermetallic compound between the two.

THE BAD SOLDERED JOINT

The condition for a good soldered joint to take place – illustrated previously in Figure 12.3 and shown photographically in Photo 12.3 – is that solder comes into contact with metal to wet it. Anything that prevents solder from coming into contact with metal, therefore, can work to prevent a good soldered joint from taking place.

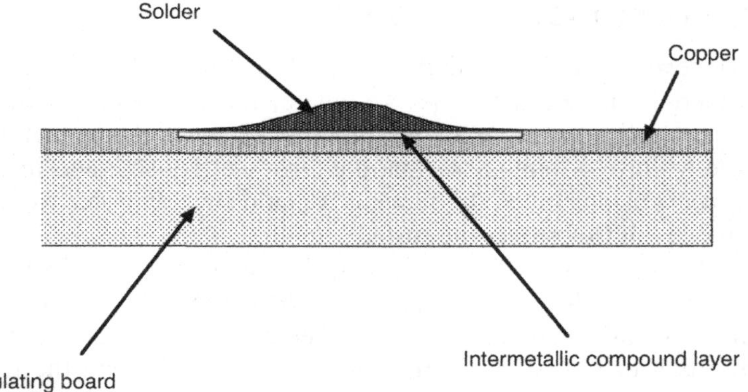

FIGURE 12.3 A cross-section of a printed circuit board, showing a single soldered joint. Note the thin intermetallic compound layer between the solder and the copper.

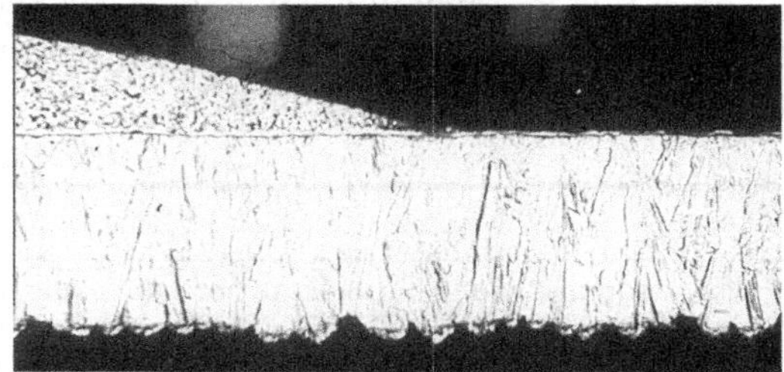

PHOTO 12.3 Micro-section photograph through an actual soldered joint, where the intermetallic compound layer between the solder (above) and the copper (below) is clearly visible.

> **Take Note**
>
> Note that we are talking about soldering joints with copper. Don't even think about trying to solder to aluminum – it just won't work. If you need to make an electrical connection to aluminum, do it with a nut and bolt!

The converse of the good soldered joint is one where the solder does not wet the joint, so does not come into contact with the metal. Such a bad soldered joint is shown in Figure 12.4.

In practice, there are several levels of wetting that do not make a good joint, and these are worth considering as they give a clue as to the problems that we can experience in soldering electronic circuits.

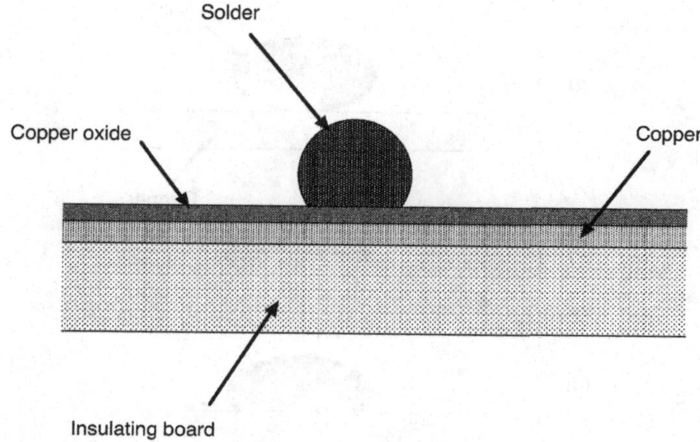

FIGURE 12.4 A bad soldered joint – where a layer of metal oxide on the surface of the metal has prevented wetting of the solder to the metal taking place.

> **Take Note**
>
> Many conditions can occur that create a bad soldered joint, and there are many types of bad soldered joints, so it is not always easy to specify exactly what has gone wrong in the soldering process. However, the usual problem is that the metal surface is not clean. Any number of contaminants may prevent a good soldered joint from taking place – dirt, grease, and metal oxide being the main ones. So – KEEP IT CLEAN! OK?

All the various levels of wetting are described below, and Figure 12.5 shows the main levels by way of visual comparison, while Photos 12.4–12.7 are photographs showing how they appear in a real-life situation.

- **Non-wetting.** Pretty obvious to the eye – solder does not flow on to the metal at all. It will not adhere to the printed circuit board's copper foil track at all. A circuit built with no wet joints will probably not work at all.
- **Partial wetting.** While not flowing and completely wetting the metal surface, a partially wet joint will appear to be solid. However, it is unreliable and may easily break off with time or vibration. Interestingly, the circuit itself may work initially when partially wet joints are present, but may not work for more than a short time.

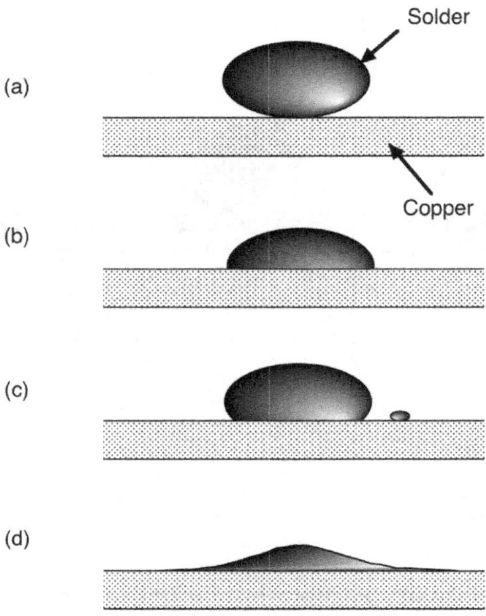

FIGURE 12.5 Four extremes of soldered joint: (a) non-wetting; (b) partial wetting; (c) de-wetting; (d) total wetting.

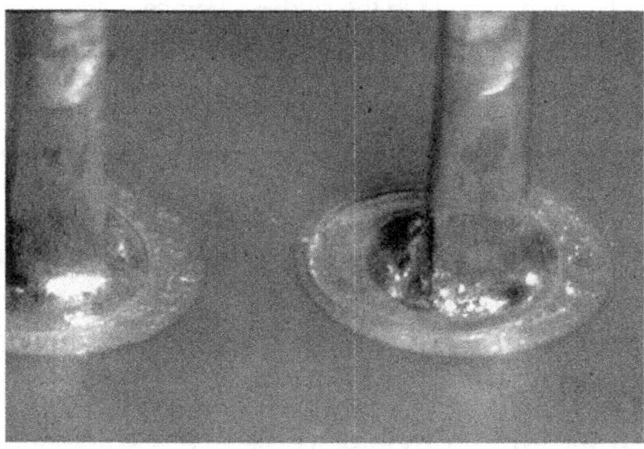

PHOTO 12.4 Non-wetting of a soldered joint – apart from a tiny area, the solder has not wet either the copper track or the component lead. Such a joint may or may not work initially, but will certainly fail before long (ECS).

- **De-wetting.** Such joints may appear to have been created properly, but in the cooling phase after heat has been removed, the solder retracts – often leaving small solder balls around the joint. As in partial wetting, the joint may initially cause a circuit to work, but will break down fairly quickly.

PHOTO 12.5 Partial wetting of a soldered joint – some of the joint's surface has been wet with solder, others haven't. The joint may work properly or may not (ECS).

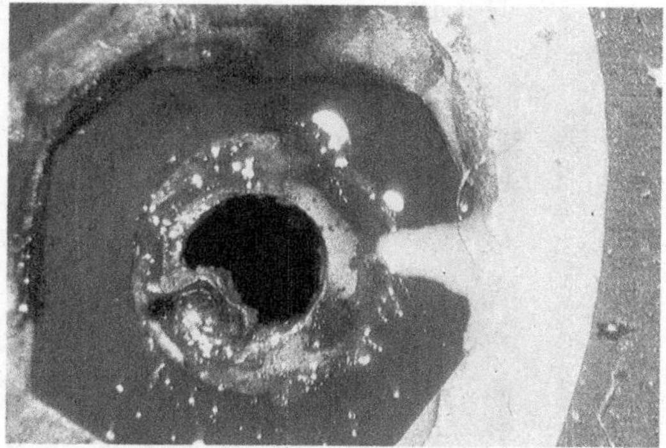

PHOTO 12.6 De-wetting of a soldered joint – solder has wet the surfaces, then withdrawn away from them, so that only a tiny area of solder remains. The joint probably does not work satisfactorily (ECS).

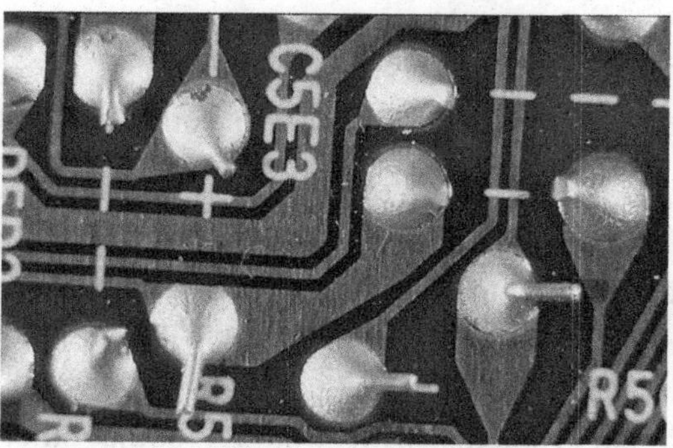

PHOTO 12.7 Total wetting of soldered joints – joints have been fully wetted by the solder making good joints (Alpha Metals).

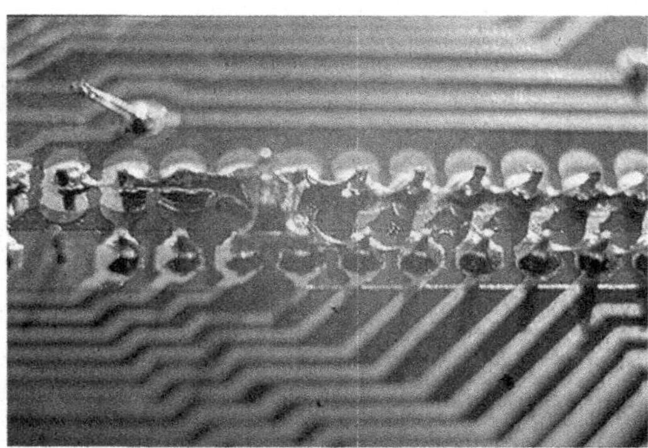

PHOTO 12.8 Bridging – too much solder has been applied, causing bridging between adjacent joints (ECS).

PHOTO 12.9 Cracked joints – the joint has probably been moved before it was cool, causing the solder to crack. This particular joint may work at the moment, but will probably fracture totally soon (ECS).

- **Total wetting.** Gives good joints, of course, which will create circuits that work and continue to do so. Note the difference between totally wet joints and all the other wetting variants. A totally wet joint is concave in appearance (and will actually be quite shiny to the naked eye). All other wetting variants of joints will be convex (if not rounded) and may be dull in appearance to the naked eye.

As well as problems with wetting, some other causes of poor soldered joints are more physical, and are shown in Photos 12.8–12.11:

PHOTO 12.10 Incomplete fillet – although wetting has taken place, there is insufficient solder to complete the joints. They may work at the moment, but will probably break down at some time in the future (ECS).

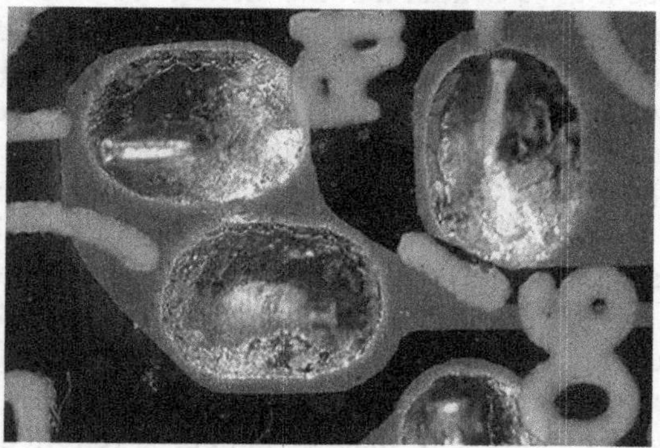

PHOTO 12.11 Residues of material have been left after the soldering process, looking unsightly and potentially causing joint failure (ECS).

- **Bridging.** If too much solder is applied, there is a risk that it may flow across the insulating board of a printed circuit board, to another part of the copper track, thus potentially causing a short circuit.
- **Cracked joints.** If the joint is moved too soon, or (if the copper track is weak or dirty) if it is moved at all, the joint may fracture, causing a broken circuit.
- **Incomplete fillet.** If not enough solder is applied to the joint it may be incomplete. While the joint may function initially, there is a potential of breakdown with time.
- **Residues.** If the joint is not fully clean before soldering, there may be harmful residues remaining after the soldering process that attack the joint over time.

Hint

As you can see here, one of the main causes (if not the main cause) of faults in electronic circuits is bad soldering. Therefore, it makes sense to be able to solder well. Practice it before you attempt to build an electronic circuit till you are able to make good joints every time. Don't bother using printed circuit boards for this, instead use a few bits of stripboard (see later for details) – even just a couple of boards give you many holes and copper strips to practice with. By the time you've soldered a couple of handfuls of components and a few short lengths of lead to the stripboards, you'll be a past master at the art of soldering!

FLUX

As we've seen, it's essential that the metal to be soldered is clean. The problem is that copper, the metal used to make copper-clad printed circuit boards that we use in electronics, rapidly oxidizes. A minutely thin surface layer of copper oxide forms very quickly on top of the copper-clad layer of the printed circuit board that makes up the copper foil track pattern. So no matter how we clean the copper surface, the oxide layer builds up before we can get a soldering iron to it. And copper oxide cannot be soldered!

Further, every time a printed circuit board is handled it picks up dirt and grease, both of which work to prevent solder from wetting the board's copper track.

The key to making sure that the copper is returned to pristine condition in order that solder can wet it is a substance called flux.

Fluxes are chemically active – that is, they work on the surface contaminants (copper oxide, dirt, grease) on the copper track of the printed circuit board to dissolve them. This occurs when the copper is heated by a soldering iron, in the first stage of the soldering process. Once the copper surface is clean, solder can be applied.

Take Note

There are many types of flux, usually judged by the level of chemical activity they have. Generally, the fluxes used in the solder used for hand soldering are of low activity. They are typically made by distilling the sap (or resin) from pine trees. The residue is then dissolved in a solvent, making it easier to apply.

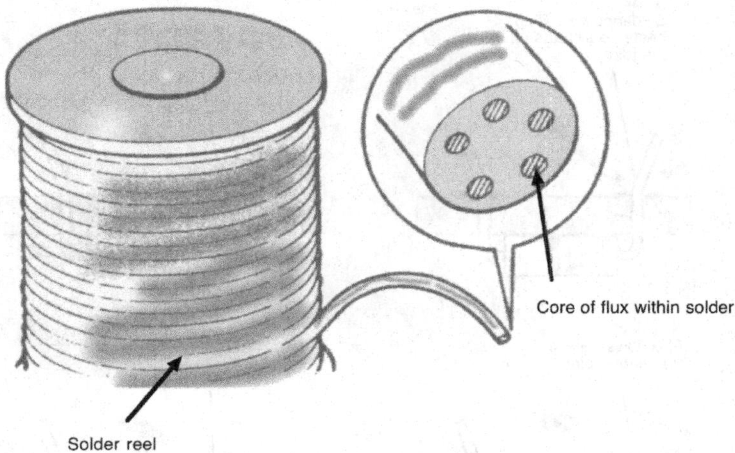

FIGURE 12.6 Cored solder, obtained in reels, appearing much like thin flexible wire. Flux is present in cores within the solder.

At first sight, it might seem that coordinating all this (heating the copper, applying the flux, waiting till the copper is clean, applying the solder) is a tricky matter, best left to the professionals and their industrial soldering processes. However, the real trick is that it can all be done by hand in a single smooth process by anyone, as long as a few simple steps are followed.

The key to the whole process is that modern solder, used to solder printed circuit boards by hand, already contains the flux needed to clean each joint to be soldered.

The solder used has cores (usually four or six) throughout its length that are filled, as it is manufactured, with flux – as illustrated in Figure 12.6. The diameter of the whole flux-cored solder lead is usually somewhere in the region of 1 mm for electronics use, though you can buy thinner diameter solder for fine work and bigger diameter for heavier jobs.

So how does it work? After all, having the flux inside the solder like this means that the flux is presented to the joint to be soldered at the same time as the solder is, rather than before it – which would suggest it has no time to clean the joint. However, the flux has a much lower melting point than the solder has, so the flux melts and flows on to the joint first. It therefore has the time to coat the metal surfaces of the joint and clean them, before the solder melts and flows over the joint to make the soldered joint, as shown in Figure 12.7. This all occurs quite quickly in a well-made soldered joint.

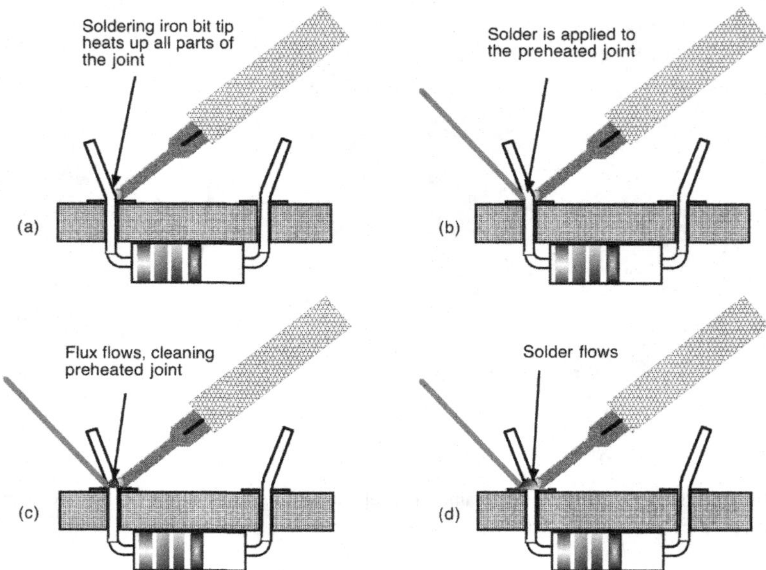

FIGURE 12.7 The main stages of hand soldering a joint: (a) the soldering iron is applied to the joint, to heat it; (b) while maintaining the soldering iron's position, cored solder is applied to the joint; (c) flux melts quickly and flows from the cored solder on to the joint's metal surfaces, to clean them and protect them; (d) the solder melts and flows over the joint's metal surfaces.

Hint

This whole process, from first applying the soldering iron to the solder flowing over the joint surfaces, actually takes only a few seconds (depending mainly on the size of the joint and how powerful the soldering iron is). If a joint takes longer than just a few seconds to make, there's a good chance the soldering iron isn't powerful enough.

Take Note

Throughout this chapter we look at processes involved in soldering various joints on printed circuit boards (i.e. those circuit boards made from copper-clad board, with copper foil tracks). The processes, however, are actually identical whatever sort of circuit boards you use.

SOLDERING IRONS

We've talked about soldering irons a little, but we've not looked closely yet at the types you can buy, or their pros and cons. That's what we are going to do now.

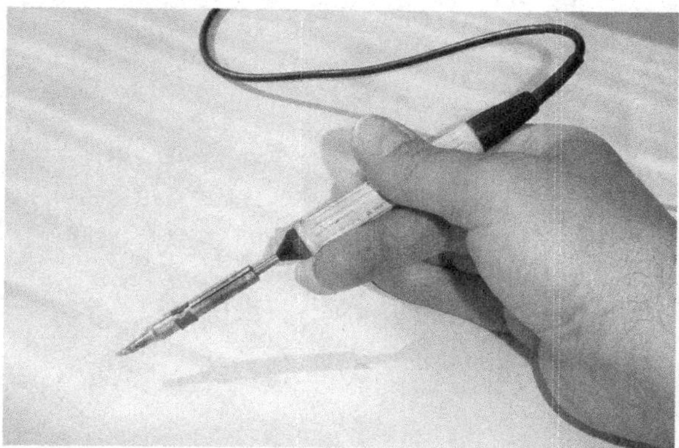

PHOTO 12.12 A mains-powered soldering iron. Note the removable bit at the left, which slides on to and off the heated element, making replacement easy.

In electronics usage terms, soldering irons don't need to be expensive, and they are priced starting at just a few pounds (the cheapest I located at the time of writing was £10 – about US $17). However, they increase in price depending on what you want, and it's easy to spend £100 (US $165) if you get carried away. So, it's probably best now to look at the main types of soldering iron available, and find out which type is best for you, and of course your budget.

First off, I should say that even the cheapest iron I found – at £10 – would do what we want from it in electronics soldering terms, so any extra expense should be carefully considered before jumping to a purchasing conclusion.

Mains-Powered Soldering Iron

The cheapest soldering irons are simple mains-powered ones. They feature a bit that does the business end when melting solder, an element that heats the bit up, and a handle that means you can pick up the iron and use it without being burned. An example is shown in Photo 12.12. This particular iron is mine, and I've owned it for many a good year (many more than I care to mention). While being fairly cheap (the modern-day equivalent of my model is around £20 – about US $35), such an iron will provide many years of perfectly adequate service on the electronics bench. You can get cheaper variants and you can get more expensive variants, but even the cheapest will do the job well.

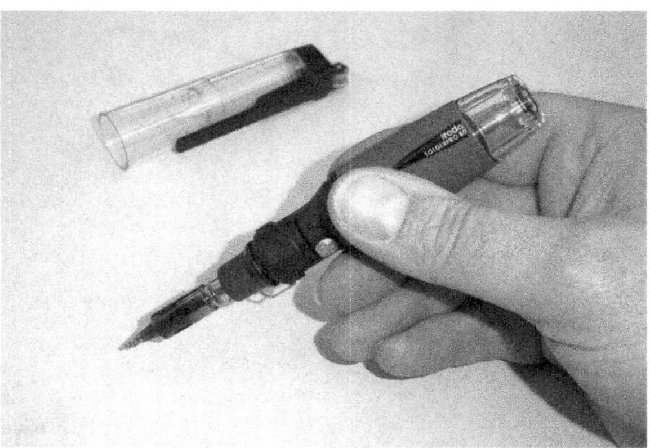

PHOTO 12.13 A gas-powered soldering iron. Butane gas is used to power the iron, and a single charge can give 20 minutes or so of use. These soldering irons are very useful where mains power is unavailable.

While a low price is the obvious advantage of such soldering irons, their main disadvantage is that bit temperature is not regulated in any way – and can vary considerably due to environmental conditions, as well as use (standing temperature may be relatively high, while working temperature – as the iron is used to make soldered joints – can be considerably lower).

Gas-Powered Soldering Irons

In a mains-powered soldering iron the heat is provided by an electrical element. An extremely useful alternative is the gas-powered soldering element, in which ordinary butane gas (such as lighter gas) is used as the heat source. An example is shown in Photo 12.13. In such soldering irons, a store of liquid gas is contained in the handle, and a burner valve is positioned close to the bit, to provide the required heat. The price of these gas-powered soldering irons is within the £20–50 (US $35–85) range.

A big advantage of gas-powered soldering irons is the speed with which the bit heats, while another is their portability – they can be used where mains power is totally unavailable, and often allow up to 20 or 30 minutes of use on a single charge of gas. The main disadvantage is the unregulated bit temperature. Gas-powered soldering irons are very useful bits of kit.

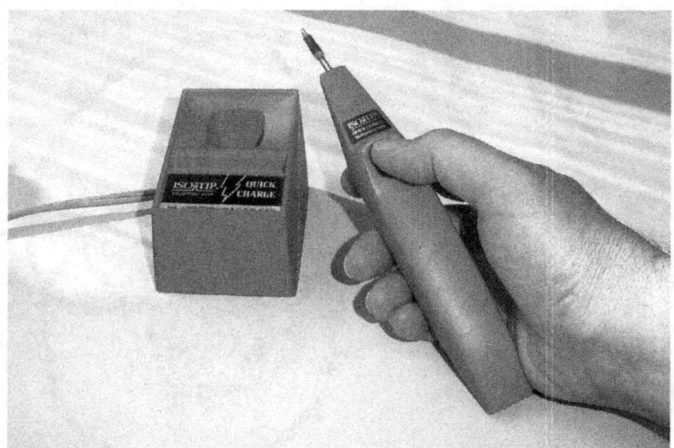

PHOTO 12.14 A rechargeable battery-powered soldering iron. To the left is the mains-powered base station, which recharges the iron's batteries when the iron is not in use.

Battery-Powered Soldering Irons

Soldering irons powered by battery are available; some – like that in Photo 12.14 – feature rechargeable batteries and a mains-powered stand or base station that recharges the batteries when not in actual use, while others use ordinary off-the-shelf cells. The element of a battery-powered soldering iron heats up the soldering bit within seconds, and it's usually only necessary to press the soldering iron button as you pick up the iron. By the time you are in position and ready to heat the joint and apply solder the bit is reaching operating heat, so battery life is usefully extended. Like gas-powered soldering irons, battery-powered irons have the advantage of portability and distance from a power supply, but the batteries will only provide power for a few minutes between recharges or replacement. Like the soldering irons discussed so far, their other main disadvantage is unregulated bit temperature. Battery-powered soldering irons using standard cells aren't expensive (£10 or £20 or so – US $17–35), while those with rechargeable batteries and mains-powered base stations are a little more (£50–100 or so – US $85–165).

Soldering Iron Stations

All the soldering irons we've considered so far here have the ultimate disadvantage of unregulated bit temperature – while

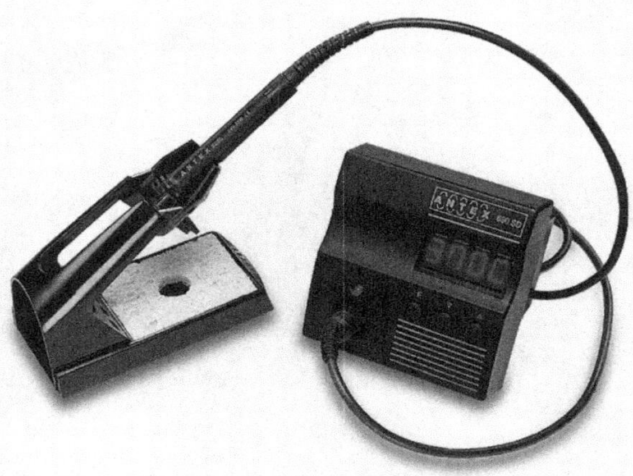

PHOTO 12.15 A soldering iron station, showing the iron itself in a stand (Antex Electronics).

standing the bit could be extremely hot, but while in use the bit temperature drops considerably. On the face of it, this isn't a massive disadvantage, as we simply work around it while we use it. However, we must be careful, because at its hottest, the bit temperature may damage some fragile components. In use, on the other hand – particularly if you solder many joints at a time – the temperature may drop sufficiently to prevent the solder flowing adequately.

Ideally, some means of controlling the bit temperature to within defined limits is required. That's where soldering iron stations are useful. They are complete units, comprising a soldering iron, a stand to hold the iron when not in use, and a controlling circuit of some description. A typical soldering iron station is shown in Photo 12.15. The controlling circuit typically controls the power to the soldering iron element, so that the bit stays at a constant, defined temperature. While such soldering iron stations can be expensive, ranging from around £40 to over £100 (about US $65–165), they are excellent tools if you do a considerable amount of soldering.

Soldering Iron Bits

Alongside the soldering iron itself, you need to think about the bit. The bit is the part of the soldering iron that comes into

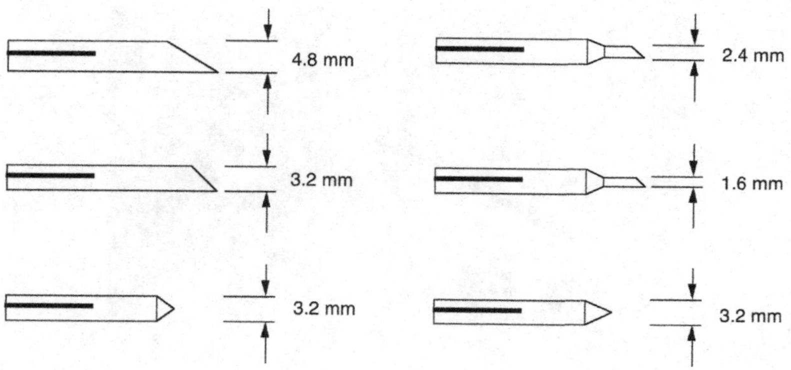

FIGURE 12.8 Typical sizes and shapes of soldering iron bits.

contact with the parts to be joined, so is a fairly critical part of the process.

A typical range of bit shapes is shown in Figure 12.8. Usually, for soldering electronic component joints an angled bit of about 2.4 mm or so is ideal.

How the bit fits on to the soldering iron element is of concern, for two main reasons:

- It determines how much of the heat generated by the iron's element passes to the bit's tip – and hence to the joint being soldered.
- It determines how easy it is to remove the bit – important if you are to solder different types of joints (for example, a bit with a smaller diameter would be used for finer joints) or for when you need to replace an old and worn bit.

In general, bits that slide over a soldering iron's element are most efficient, allowing much of the heat generated by the element to pass to the bit's tip. Photo 12.16 shows such a bit, taken off its element.

Bits that fit inside an element, or screw on to it, are probably best avoided, if only because the very nature of the soldering process means that metal bits often corrode – even slightly – due to heat, and this may mean that the bit becomes hard to remove. Excessive pulling or twisting of the bit to remove it may then damage the element.

Some soldering irons use a bit type that also comprises the element. The bit/element combination attaches to the soldering iron

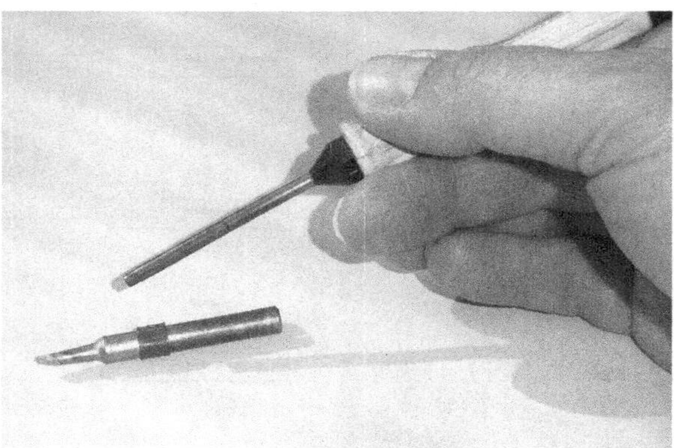

PHOTO 12.16 A soldering iron whose bit slides on or off the element. The bit is easily removed and/or replaced, and provides an efficient heat transfer between the element and the bit's tip.

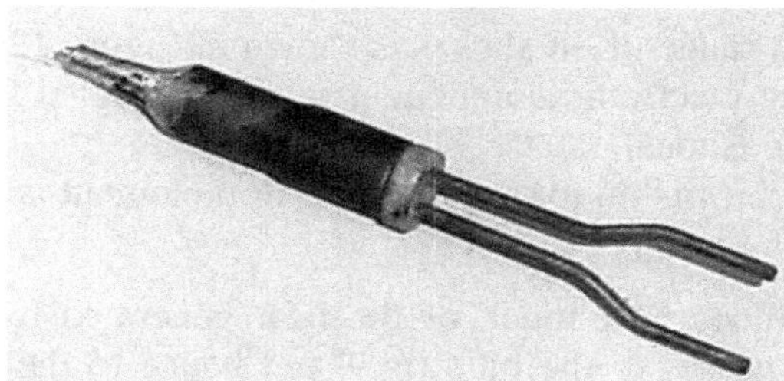

PHOTO 12.17 A unified element/bit that fits to the soldering iron's body by pushing in the two prongs.

by two prongs that slide into the soldering iron body. Element and bit tip being unified in this way, there is a highly efficient heat transfer between element and tip, and this design has the great advantage that changing the bit effectively replaces the element in one operation. An example of such a tip is shown in Photo 12.17.

SOLDERING IRON ACCESSORIES

Soldering Iron Stand

There's a few accessories for soldering irons that are worth buying. First – and most important – of these is a soldering iron stand (a typical stand is shown in Photo 12.18), which holds an iron while it's not in use. Soldering irons, of course, by their very

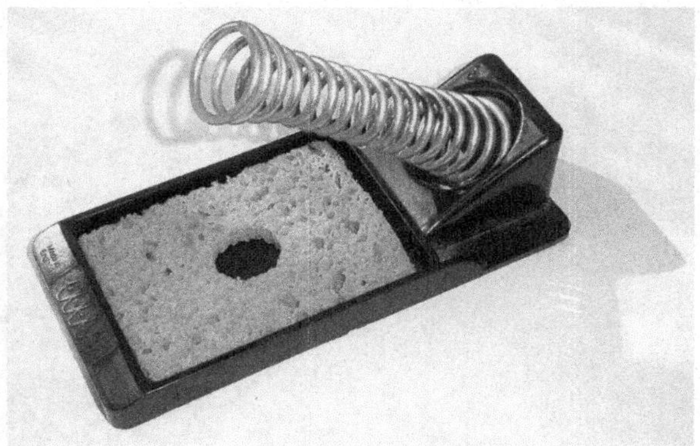

PHOTO 12.18 A soldering iron stand. This particular stand features a sponge.

> **Take Note**
>
> As you'll see soon, a clean soldering iron bit tip is one of the prerequisites of good soldering. Put another way, if your soldering iron bit tip is not clean, don't expect to make good soldered joints, and do expect your circuit not to work!

nature are hot, so having a stand to put one in when not actually using it is quite a safety feature. The stand in Photo 12.18 is around £7 or so (about US $12) – so it shouldn't break the bank.

Sponge

Talking of a soldering iron stand with a sponge, a sponge is actually a very handy accessory to have. Even if you have to resort to – as I have done on occasions – using a simple kitchen sponge, dampened with water, stood on a teaplate, then so be it!

The £7 soldering iron stand of Photo 12.18, featuring an integrated sponge, is a great way to combine the two accessories – not essential, just neat.

You use a dampened sponge to wipe your hot soldering iron bit tip on, immediately prior to tinning it (see later) and using it to solder joints.

Tip Tinning and Cleaning Block

Another way to keep your soldering iron bit's tip clean is to use a tip tinning and cleaning block. It's a thin disk-shaped block in

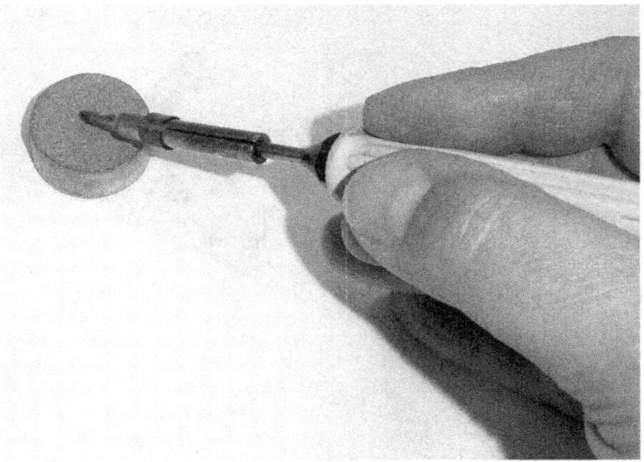

PHOTO 12.19 Using a tip tinning and cleaning block to er … clean and tin a hot soldering iron bit tip.

a metal container that can be carried in a pocket, or mounted on a suitable surface close to the soldering iron. In use, you simply wipe the hot soldering iron bit's tip over the block's surface, simultaneously cleaning and tinning the tip, as shown in Photo 12.19. It's convenient, it's handy, it's cheap (around £10 – US $17), and it saves having a wet sponge on your worktop.

KEY PRACTICAL POINTS WHEN SOLDERING

So far, we've just looked at the theoretical aspects of solder and soldering. We've not once really considered how you actually use solder to create a soldered joint – we've merely considered the theory behind the practice. But this book isn't only about the theory of electronics, it is also about the practical aspects. Now, for the first time in this chapter, we are about to consider exactly how to solder, in a highly practical way – in other words, we're going to follow step by step the soldering procedure, so that you, the reader, can attempt soldering as we follow the book. We'll also look at the tools you need, and the components and parts you may come across while soldering a project of your own. OK, enough said, let's look at one last little bit of theory, then we'll get started.

The whole soldering process in electronics depends on six important aspects, summarized now. When soldering:

Hint

OK, this is it! We have had all the theory about soldering, but from here we look at the practical side of soldering. The key to it from now is in your hands – PRACTICE, PRACTICE, PRACTICE. The more you do, the better you get. When you start, your soldered joints will be hit or miss. But as you practice they will improve. That's a promise!

Take Note

You'll see in Photo 12.20 that a real printed circuit board is being used. This is, as we've already seen, a thin insulating board with a copper pattern or track on the surface. Such printed circuit boards are available in kit form complete with components from electronics suppliers, or if you are building up a project from an electronics magazine, say, may be available from the magazine publisher. The more experienced enthusiast may also like to make their own printed circuit board. But such a printed circuit board is not always necessary. Occasionally, particularly if the project is a fairly simple one, another type of printed circuit board may be used: where the track is not specific to the circuit, but is a simple matrix of copper tracks, running parallel along the board's surface, with regularly spaced holes (usually at 0.1 inch distances). This type of printed circuit board (Photo 12.21) is known typically as a stripboard, and can be bought in several sizes. Photo 12.22 shows such a stripboard being cleaned.

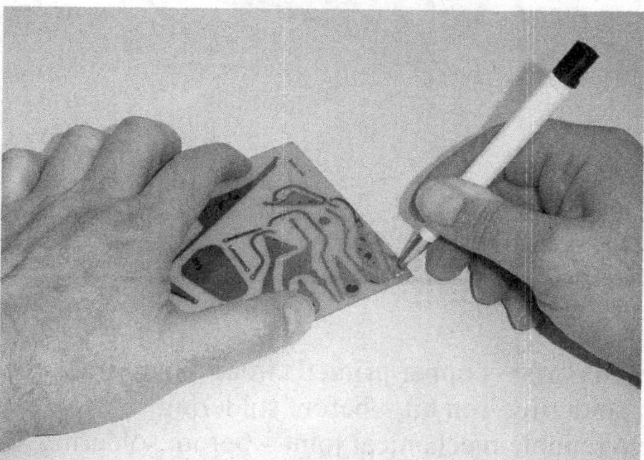

PHOTO 12.20 Cleaning the copper track of a printed circuit board using an abrasive fiberglass pencil – simply rub the pencil over the track, particularly over the component land areas, to clean it. Brush aside the fiberglass residues before commencing soldering.

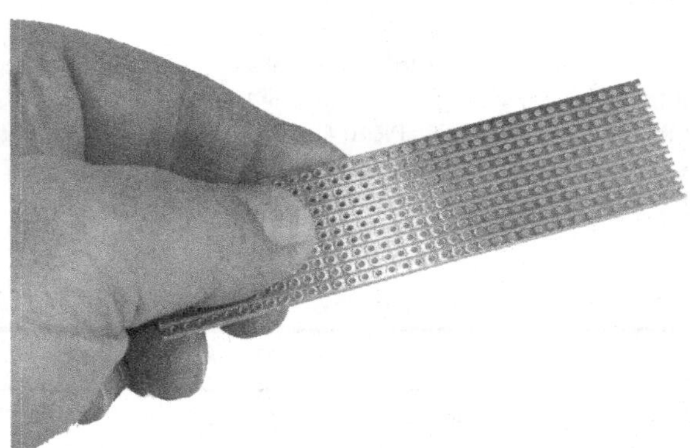

PHOTO 12.21 Stripboard. Note the parallel tracks of copper and the regular intervals of holes (a total of 380 in this small piece of stripboard). Such a printed circuit board allows a quick method of soldering components into a circuit, albeit not perhaps as neatly as a purpose-designed traditional printed circuit board.

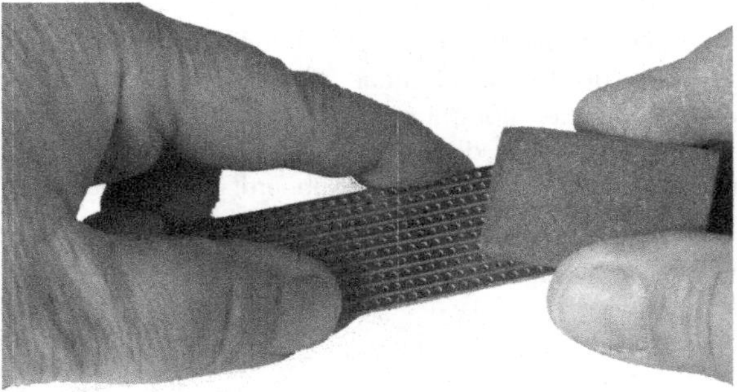

PHOTO 12.22 Cleaning the parallel tracks of a stripboard printed circuit board using an abrasive scrubbing block.

1. Clean all parts – copper printed circuit board track, component leads, soldering iron bit – before soldering.
2. Make a reliable mechanical joint – before soldering.
3. Heat the joint sufficiently – before applying solder.
4. Apply the solder – keeping the soldering iron on the joint while the solder melts.
5. Remove the soldering iron and allow the solder to solidify – before handling or moving the joint.
6. Trim the excess component lead from the joint, once the joint has solidified and cooled sufficiently to allow handling.

Hint

For practice at least, even if you don't intend to use it again, stripboard printed circuit board is a cheap and ideal board to use to get your soldering up to speed. Even just a small board a few centimeters square will give you several tens if not hundreds of holes for you to solder component leads into. You can keep practicing until your soldering is top notch, and you won't have spent a fortune getting there.

Let's now consider these six aspects in detail, and get some hands-on experience of soldering.

Cleaning All Parts

We already know that flux is used automatically when soldering using cored solder to help clean the joint surfaces, as part of the soldering process. However, flux can only remove so much oxide, grease, or dirt. It's best to improve your chances of making a good solder joint by manually cleaning the joint surfaces first. Use an abrasive fiberglass pencil (as shown in Photo 12.20) or an abrasive scrubbing block (see Photos 12.22 and 12.23), rubbing them over both copper track and component leads to minimize dirt and grease.

Not just the copper track of the printed circuit board needs to be cleaned – the components that are to be soldered into position must be clean too. The easiest way to ensure this is to use either an abrasive fiberglass pencil or an abrasive block to clean the component leads, as shown in Photo 12.23.

It's vitally important that the soldering iron bit tip be kept clean too, and the best way to do this is to tin it. Tinning is a process whereby fresh solder is regularly melted on to the heated bit, so that the bit remains coated with fresh solder. Flux-cored solder should be used, just as when soldering the actual joint, as the flux within the cores cleans the bit simultaneously.

The process is a two-stage one: first wipe old excess solder off the soldering iron bit tip, then tin the bit by applying fresh solder. Photo 12.24 shows a wet sponge being used to wipe the soldering iron bit on. An alternative, shown before in Photo 12.19 but included again here for completeness, is shown in Photo 12.25, where a bit cleaning and tinning block is being used to keep the soldering iron bit tip clean immediately prior to tinning it.

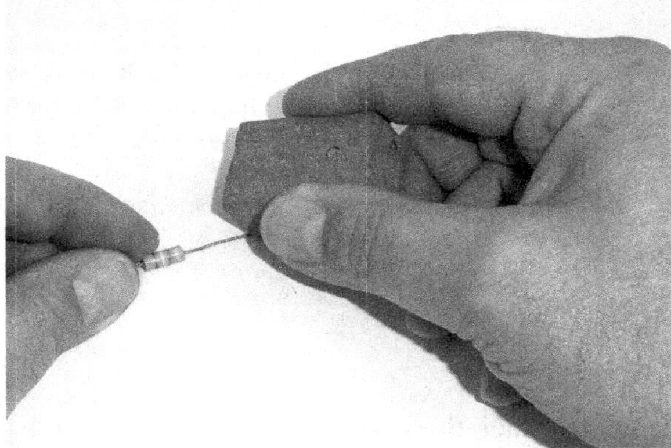

PHOTO 12.23 Using an abrasive scrubbing block to make sure a component's leads are clean, prior to soldering.

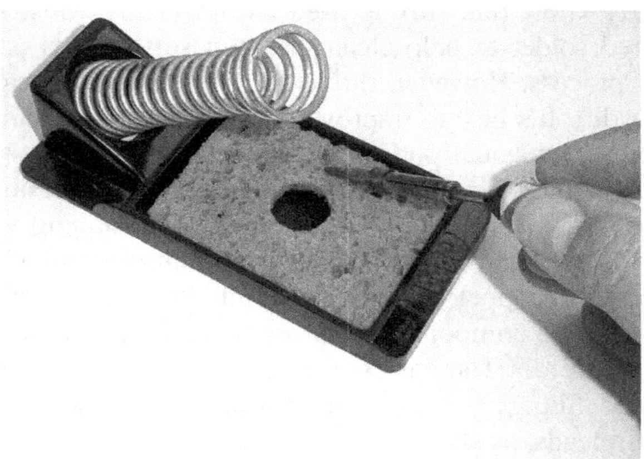

PHOTO 12.24 Cleaning a soldering iron bit tip using a wet sponge. Simply wipe the tip over the sponge surface.

Hint

If you have bought yourself a new soldering iron and are preparing to tin it, you will find that when you first plug the soldering iron in to turn it on, smoke will be created and given off from the bit tip – this is because the bit is coated with a grease substance to prevent it oxidizing. This is useful, but also means that as the bit heats and burns off the grease, oxide very quickly forms, making the bit unusable. To prevent this, tin the bit tip as shown in Photo 12.26 immediately as it heats.

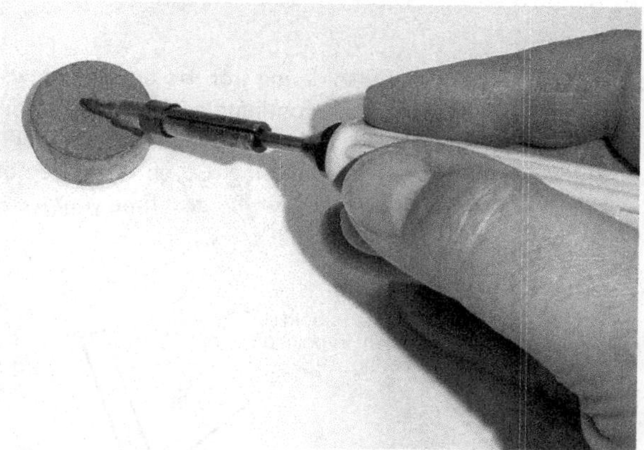

PHOTO 12.25 Using a tip cleaning and tinning block to clean a soldering iron bit tip – just wipe the tip over the block.

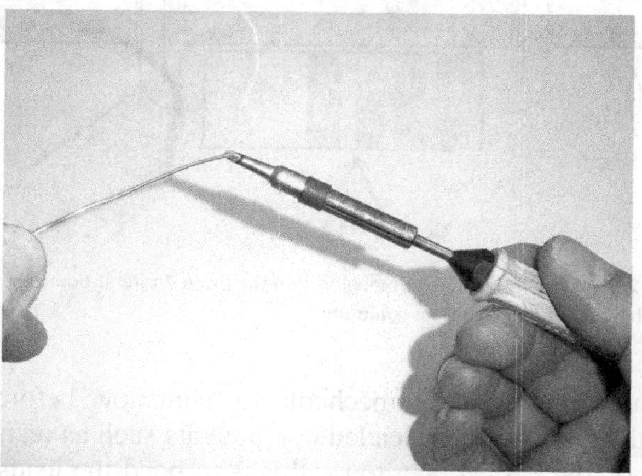

PHOTO 12.26 Tinning a soldering iron bit tip. Once the soldering iron bit tip has been cleaned as shown in either Photo 12.24 or Photo 12.25, quickly tin it using a small amount of ordinary flux-cored solder. The tip will have a shiny appearance if tinned properly.

Immediately following removal of old excess solder, apply fresh solder to the soldering iron bit tip, as shown in Photo 12.26. Just a small amount of solder is all that's necessary to keep the tip in err … —tip-top condition.

Making a Reliable Joint

Components must not move while you are soldering a joint, otherwise the joint may be faulty. As a general rule, it's best to fix

Hint

Get into the habit of tinning the soldering iron bit tip every few minutes while it is on, to maintain it in peak condition and ready to solder joints. Conversely, if you're not going to use the soldering iron for a while, turn it off – that way you don't have to keep tinning it even while it's not in use, and it will stay in better condition ready for the next time you need it.

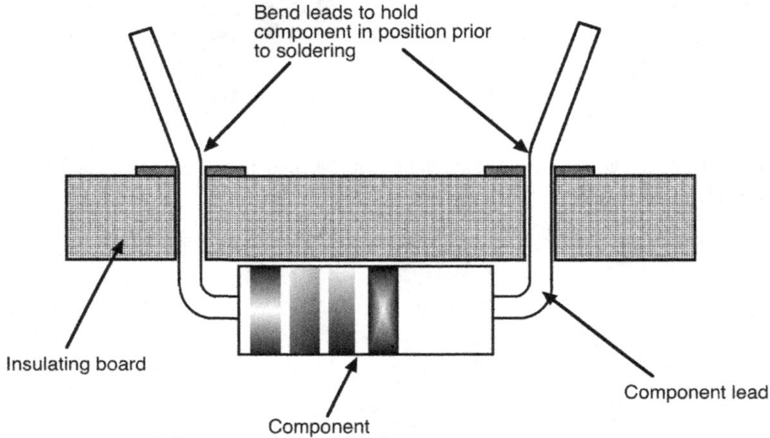

FIGURE 12.9 After inserting a component into the printed circuit board, bend its leads slightly to hold it in place ready for soldering.

the components in place mechanically somehow before soldering. With axial- or radial-leaded components such as resistors and capacitors, the easiest way to do this is to bend the leads slightly after insertion into the printed circuit board, as shown in Figure 12.9. As a guide, the angle of the bend doesn't need to be (and shouldn't be) any more than just a few degrees – bend the leads only as far as necessary to hold the component in place.

Heat the Joint

For successful soldering, all the metals forming the soldered joint should be preheated. Preheating is easily accomplished – apply the soldering iron bit tip to the joint – touching both the copper track and the component lead, as shown in Photo 12.27. The time you should preheat the joint for depends on the size of metal to be heated and the power of the soldering iron, but generally this

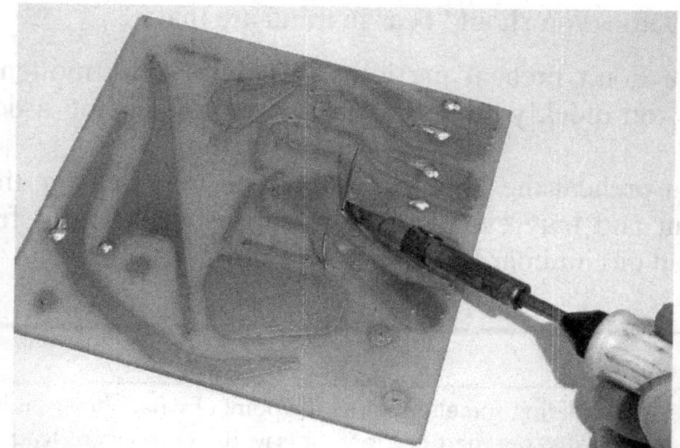

PHOTO 12.27 Preheating the joint to be soldered – apply the soldering iron bit tip to the joint's metal surfaces.

Take Note

If you insert more than one component into a printed circuit board at a time before soldering, and bend their leads outwards to prevent them falling out, be careful that when you solder their joints solder doesn't bridge between them, as shown in Figure 12.10. When angled leads like this are soldered, the joint contains much more solder than a lead that is perpendicular to the printed circuit board.

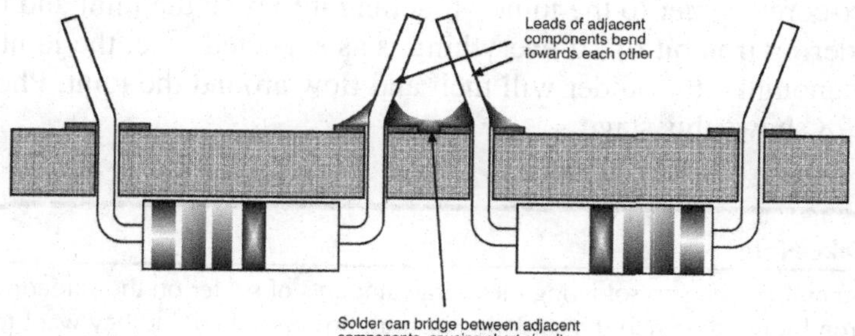

FIGURE 12.10 How solder can easily bridge between joints if angled component leads are used to hold components in place prior to soldering – take care when using this method to prevent such short circuits.

will be for no more than a few seconds – between about 2 and 8 seconds, say. You will learn to judge the preheat time required by joints with practice, but in any case it's not normally that crucial.

The factors you should bear in mind are that:

- If you don't preheat the joint sufficiently, the molten solder cools too quickly and you run the risk of making a defective joint.
- If you preheat the joint too much, you may damage the component and may even cause the copper track to lift from the printed circuit board.

Take Note

Beginners to soldering sometimes miss the point of what's been said, and – wrongly – heat just one part of the joint (say, the component lead poking through the printed circuit board). Sure, when the solder is then applied the solder will melt, but it will melt as a blob that doesn't properly flow on to the remainder of the joint (that is, the copper track). It may look like quite a successful joint, but because part of the joint is relatively cold when the molten solder flows across it, the joint will be weak, incomplete, or unreliable. You must make sure that the soldering iron bit tip touches all parts of the joint, so it all heats up to a sufficient temperature before applying solder.

Apply the Solder

With the soldering iron bit tip in position, apply the end of the flux-cored solder to the joint – touching it to both the joint and the soldering iron bit tip. If everything is as expected – i.e. the joint is hot enough – the solder will melt and flow around the joint. Photo 12.28 shows this stage.

Take Note

Some beginners to soldering melt large amounts of solder on the soldering iron bit tip, then dab the soldering iron bit tip on to the joint they want to make – the soldering equivalent of putting glue on to a spatula then wiping the spatula on to something to spread the glue over it.

DO NOT UNDER ANY CIRCUMSTANCES DO THIS.

While the process might work well for glue, it certainly doesn't work for solder. Soldering is only effective when the joint is preheated, and fresh flux is applied at the same time as solder. If you put solder on the soldering iron bit tip, the flux burns away before applying it to the parts of the joint, and the joint is too cool anyway – causing only bad soldered joints.

The moral is: only use the method described earlier to solder joints!

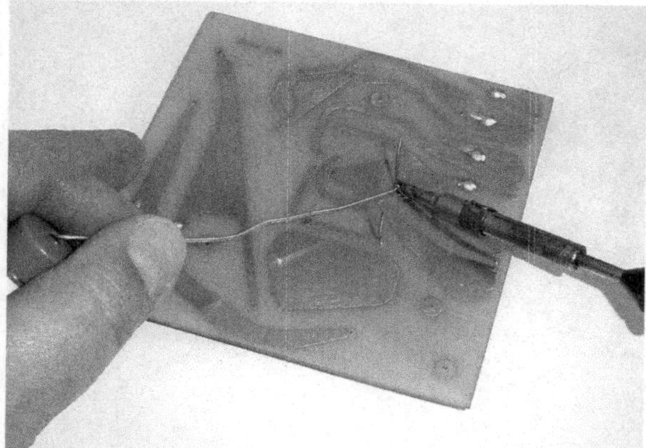

PHOTO 12.28 Solder melting and flowing around the preheated joint.

> **Take Note**
>
> A good soldered joint has a concave, shiny appearance. If the joint is not heated sufficiently, or if you move the joint and damage it, you will probably notice that it is not concave, or that it is a dull, gray color. Don't worry – just repair the fault by heating the joint again, and perhaps applying a little more solder so its flux can clean the joint, until the solder flows properly.

Remove the Soldering Iron

Once the solder has flowed around the joint (not before), quickly remove the soldering iron. Do not move the printed circuit board, and do not move the component or its leads, until the solder has solidified. If you move anything before the solder solidifies, the joint may be damaged.

Trim Excess Component Leads

If the component leads were left as they are, and more than just a couple of components were inserted into the printed circuit board, you run the imminent risk of component leads touching – effectively forming short circuits. To prevent this you have to trim all excess component leads off immediately above the joint. Photo 12.29 shows this.

After all joints on the printed circuit board have been soldered and excess leads trimmed, it's a good time to check all the joints

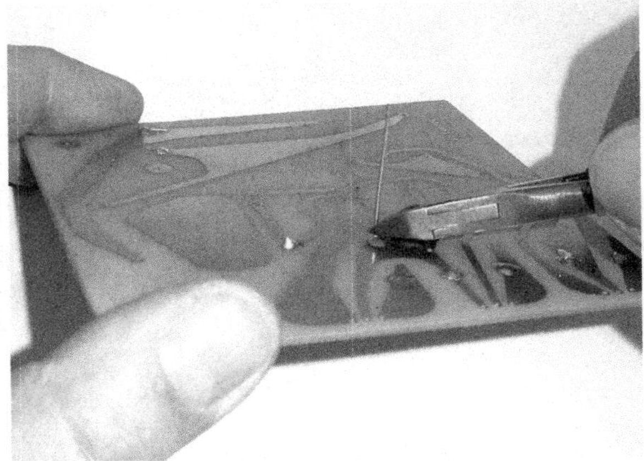

PHOTO 12.29 Trimming excess component leads off, close to the soldered joint, to prevent short circuits.

visually to see if they look "good" or "bad", repairing any that are suspect. Also check over all the copper foil track, to see that no solder bridges have formed between tracks or component joints. If you find any solder bridges, use desoldering tools – a solder sucker, or desolder braid (see later) – to remove excess solder, and resolder any joints that need it.

CONNECTING LEADS

Inevitably, any printed circuit board holding its electronic circuit is not isolated from the world around it! It must be connected – to switches, controls, off-board components, input and output sockets, and so on. So, it's common – indeed, normal – to need to solder connecting leads to the printed circuit board copper foil track.

Connecting leads are of two types:

- Single-stranded – sometimes called bell wire – where the lead is just one piece of wire, of a particular diameter. A typical connecting lead of this type would be classified as, say, 1/0.6 (which means it is one strand of diameter 0.6 mm). This type of connecting lead can be used in a situation where the wire is fixed, and not moved in any way – because constant movement may fracture the single strand of wire.
- Multi-stranded – where the lead is made up from several strands of thinner wire, together making up the particular diameter. A typical connecting lead of this type would be

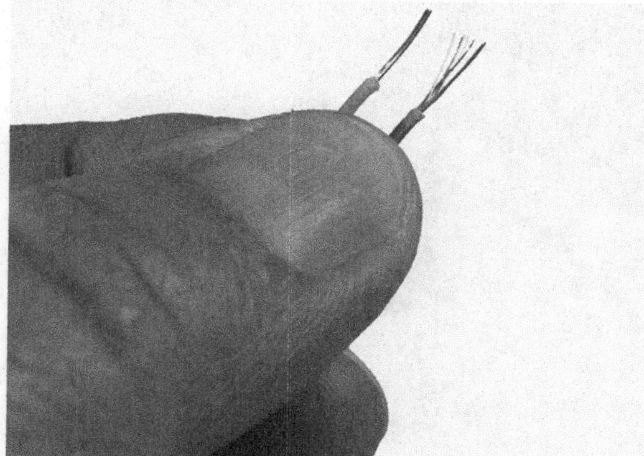

PHOTO 12.30 The two basic types of connecting leads – single-strand and multi-strand.

classified as, say, 7/0.2 (which means it is formed from seven strands of wire of diameter 0.2 mm each). This type of connecting lead is best suited for situations where there may be some movement of leads after a project has been finished – the multi-strand nature means individual strands are less likely to fracture than a single-strand lead.

Both types of lead are readily available in a selection of color insulations from most electronics suppliers. Photo 12.30 shows both types, with some insulation stripped away from the ends.

To solder connecting leads to a printed circuit board, use the following procedure:

- Strip around 5 mm of insulation from the end of the lead.
- If the connecting lead is multi-stranded, twist the loose ends of the stripped lead. If you solder the lead untwisted to the board, there's a possibility a loose strand may bridge across to another track.
- Clean and tin the soldering iron bit tip. You should always do this before any soldering operation.
- Tin the connecting lead. This has two purposes: first, it removes any oxides, dirt, and grease on the wires making up the leads; second, the solder you add in the tinning stage will be used when the joint is made – meaning you won't have to add much solder at that time. The soldering iron bit tip needs

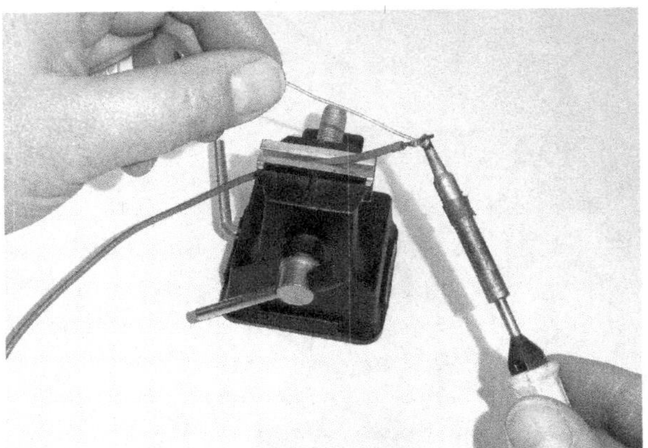

PHOTO 12.31 Tinning a connecting lead – do not heat the lead for too long, as insulation may melt or deteriorate.

to be quite a broad one – around 3 mm – for this job. To tin the lead:

- Hold the flat part of the soldering iron bit tip upwards (as shown in Photo 12.31) and lay the stripped end of the connecting wire on the tip's flat surface.
- Apply the end of a length of solder to the wire (not the soldering iron bit tip) – this means you can see when the wire is hot enough, as the solder begins to melt.
- When enough solder has flowed over the lead ends to fully cover the wire, remove the soldering iron and solder.
- Allow the lead to cool before touching or moving it.

- Tin the printed circuit board. At the point where the tinned lead is to be connected to the printed circuit board's copper foil track, the copper must be tinned, to remove oxide, dirt and grease, and to apply solder ready for the joint to be made between board and lead. To tin the printed circuit board track:
  - Clean and tin the soldering iron bit tip.
  - Apply the soldering iron bit tip to the copper track of the printed circuit board, at the point where the lead is to be connected – as shown in Photo 12.32 – and wait for a few seconds.
  - Apply the end of the flux-cored solder to the track, a small distance (about 2–3 mm) away from the soldering iron bit tip – wait for the solder to melt and flow towards the soldering iron bit tip.

> **Take Note**
>
> While it's correct to tin the printed circuit board copper foil track at the points where connecting leads are to be joined, this is not the procedure where components are to be fitted to the printed circuit board – if the copper foil track is tinned where a component lead is to be inserted, the hole for the lead may be closed with solder.

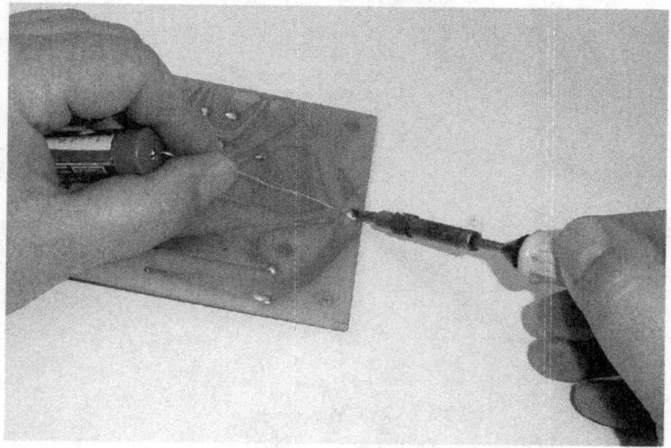

PHOTO 12.32 Tinning the copper foil track of a printed circuit board – apply the soldering iron bit tip first, and wait for a few seconds before applying solder.

- ❑ Remove the soldering iron bit tip and the solder.
- ❑ Allow the printed circuit board to cool before moving or touching the solder.

Once the lead is tinned, it can be soldered to the copper foil track of a printed circuit board. There are three main ways this can be done, soldering the lead:

1. Directly to the printed circuit board copper foil track.
2. To a terminal pin, inserted through a hole in the copper foil track.
3. Inserting the lead through a hole, purposely designed into the printed circuit board's copper foil track.

The procedures you should follow for all three ways are detailed below.

Soldering a Connecting Lead Directly to Copper Track

- Tin the soldering iron bit tip, as described earlier.
- Tin the connecting lead, as described before.

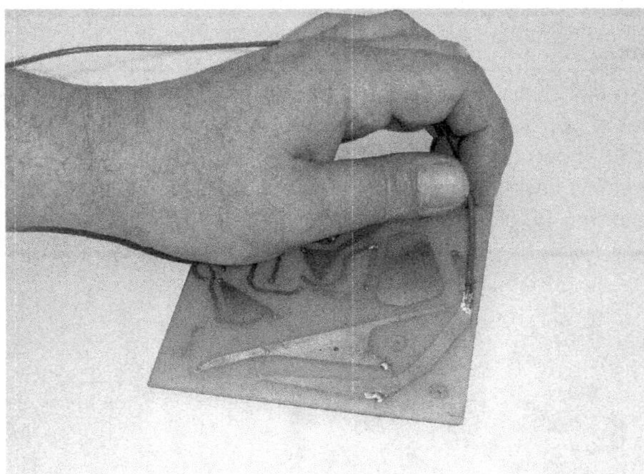

PHOTO 12.33 Position the tinned connecting lead end over the tinned copper track.

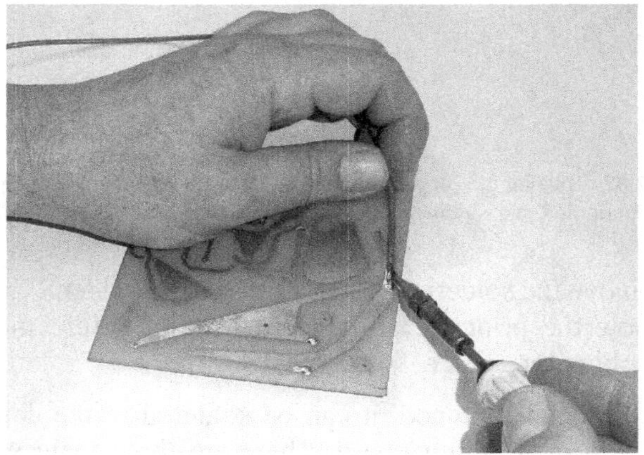

PHOTO 12.34 Heating the connecting lead end and copper track – the solder already present in the tinning on both the lead end and the track will be sufficient to make the soldered joint.

- Tin the printed circuit board copper track, at the point where the lead is to be connected, as described before.
- Hold the tinned end of the lead on the tinned copper track, as shown in Photo 12.33.
- Apply the soldering iron bit tip to the tinned lead end (not the copper track), pressing down on the lead end to hold it in position, and wait for the solder on both the lead end and the copper track to melt, as shown in Photo 12.34.

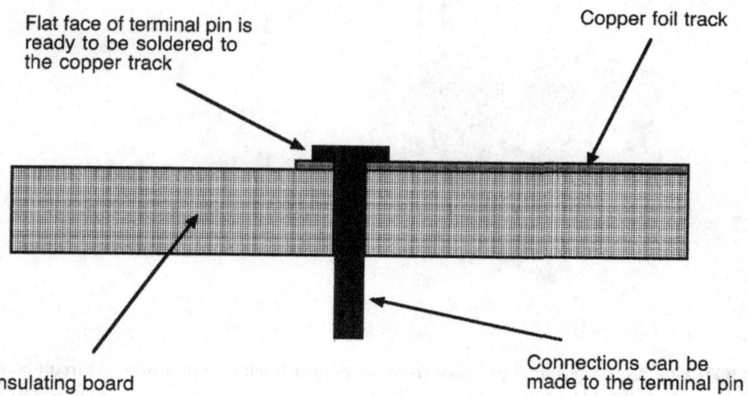
Flat face of terminal pin is
ready to be soldered to
the copper track
Copper foil track
Insulating board
Connections can be
made to the terminal pin

FIGURE 12.11 A terminal pin, inserted through a printed circuit board from the copper track side. The pin is then soldered to the copper track and a connecting wire soldered to the other end.

- Remove the soldering iron, but hold the lead without moving it.
- Wait for the solder to solidify before moving the printed circuit board or letting go of the lead.

Soldering a Connecting Lead to a Terminal Pin

Terminal pins are useful connection devices when it comes to soldering and printed circuit boards. They are short, rigid pins that are inserted through holes in the printed circuit board to allow connections to be made to the board easily. Use of a terminal pin is shown in Figure 12.11. The pin usually has a flat face that is soldered to the copper track of the printed circuit board, while the connection (typically a wire) is made to the other end of the pin, on the component side of the printed circuit board.

Using and Soldering Terminal Pins

- Insert a terminal pin into the printed circuit board – note that the pin goes through from the copper foil track side of the printed circuit board to the component side of the board, i.e. the connecting lead will be soldered to the component side not the copper track side (see Photo 12.35).
- Clean and tin the soldering iron bit tip, as described previously.
- Tin the connecting lead, as described previously.
- Apply the soldering iron bit tip to the terminal pin and wait for a few seconds (see Photo 12.36).

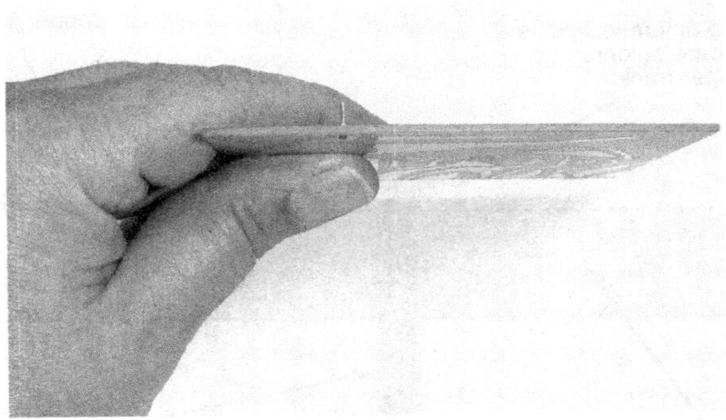

PHOTO 12.35 A terminal pin goes from the copper track side of a printed circuit board to the component side.

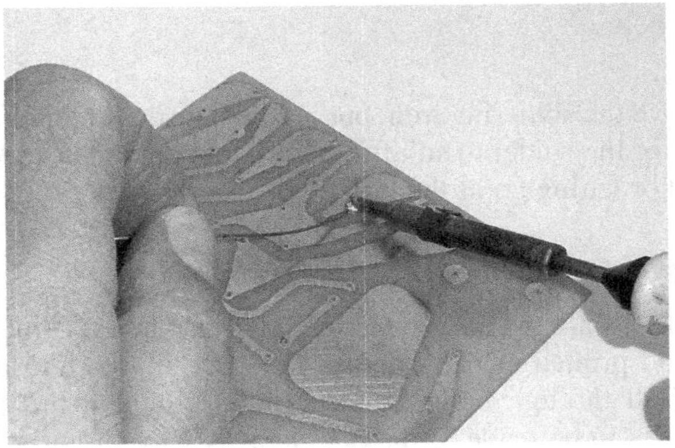

PHOTO 12.36 Soldering a terminal pin to a printed circuit board.

- Apply solder to the terminal pin.
- When the solder melts and flows over the terminal pin and the copper track, remove the soldering iron and wait for the soldered joint to cool.
- Turn the printed circuit board over, then tin the terminal pin.
- Hold the tinned end of the connecting lead against the tinned terminal pin, as shown in Photo 12.37.
- Apply the soldering iron bit tip to the connecting lead and terminal pin.
- Wait for a few seconds for the solder on both the lead and pin to melt, then remove the soldering iron.
- Do not move either connecting lead or printed circuit board until the solder has solidified.

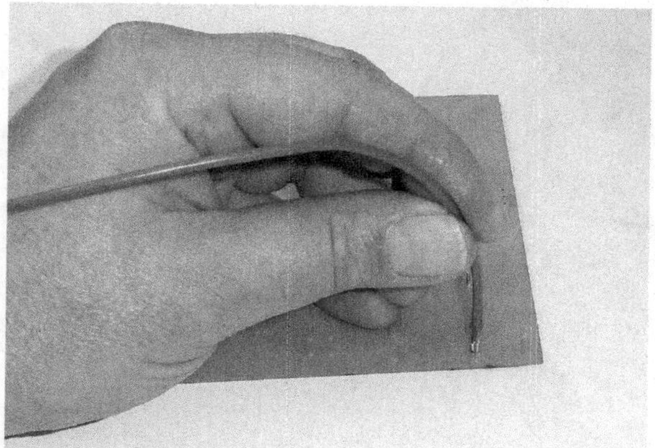

PHOTO 12.37 The connecting lead should be vertically aligned with the terminal pin, and touching it.

Hint

For both these methods of soldering a connecting lead to a printed circuit board, be extra careful not to allow the connecting lead end to move at all once the joint has been soldered and during the cooling phase, or a defective joint may result.

Soldering a Lead to a Hole

- Clean and tin the soldering iron bit tip.
- Insert the connecting lead through the hole in the printed circuit board – do not tin the connecting lead first, or it will not go through the hole!
- Apply the soldering iron bit tip to the joint, touching both the copper foil track of the printed circuit board and the lead, as shown in Photo 12.38.
- Apply solder to the joint — this tins the lead and solders the lead to the board simultaneously.
- When the solder melts and flows over the connection lead and the copper track, remove the soldering iron and wait for the soldered joint to cool.
- Do not move either connecting lead or printed circuit board until the solder has solidified.

UNSOLDERING

There are times when it's necessary to unsolder components from a printed circuit board. You may, perhaps, have inserted

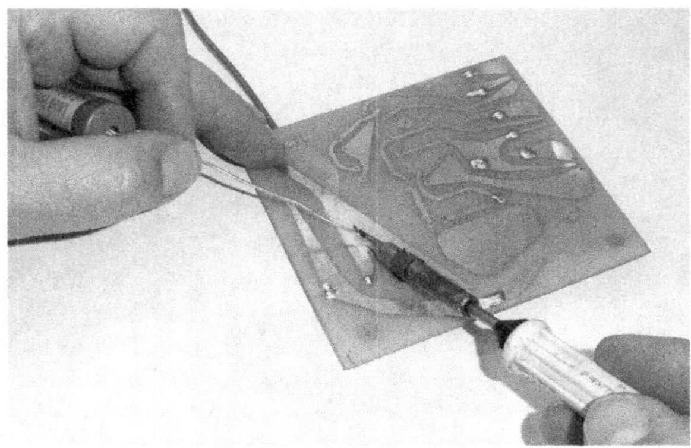

PHOTO 12.38 Soldering a connection lead to a printed circuit board.

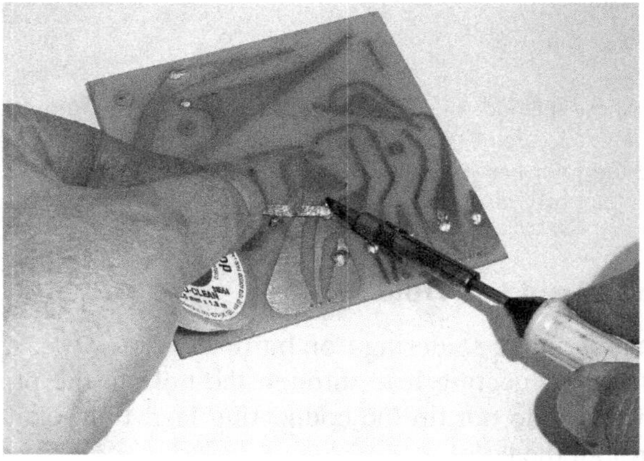

PHOTO 12.39 Desoldering using solder braid – the molten solder wicks into the strands of the solder braid away from the joint.

and soldered a component incorrectly into place, or a particular component may be faulty. In such cases, unsoldering is the only option.

There are two basic methods of unsoldering components, and both rely on having extra tools: either solder braid or a desoldering pump.

Using Solder Braid

- Tin the soldering iron bit tip.
- Place the end of the solder braid over the joint to be desoldered, as shown in Photo 12.39.

Hint

As you use a portion of solder braid to unsolder a joint, it will become loaded with solder and so cease to be usable. Using a pair of side-cutters, trim off the used portion to maintain a fresh solder braid end.

Hint

Sometimes not all solder can be removed from a joint (very often a small amount that cannot be removed will still hold the component lead to the copper track). If this occurs, you will find that you have to heat the joint with the soldering iron bit tip one last time, before quickly – before the solder solidifies – releasing the component.

- Apply the soldering iron bit tip on to the solder braid. Press gently but firmly against the joint. After a few seconds or so, the solder will melt, and some will wick into the solder braid.
- Remove the soldering iron bit tip. Place a fresh portion of solder braid on to the solder joint and repeat the last step until all the solder forming the joint has been removed.
- Repeat the process for all component joints.
- Remove the component.

Using a Desoldering Pump

- Tin the soldering iron bit tip.
- Prime the desoldering pump, by pressing the charger down, as shown in Photo 12.40.
- Apply the soldering iron bit tip to the soldered joint and wait for the solder to melt.
- Apply the nozzle of the desoldering pump to the molten solder, as shown in Photo 12.41.
- Push the pump's release button. The vacuum will suck the molten solder away from the joint and into the desoldering pump chamber, where it will solidify.
- Repeat the previous four steps as required, until all the solder has been removed from the joint.

Take Note

Every now and again, the desoldering pump chamber will fill with solidified solder to the point that it doesn't work efficiently. Empty it by repeatedly priming and discharging the pump, so that the solder is forced out of the nozzle.

Take Note

Unsoldering usually requires considerably more application of heat than soldering – the soldering iron bit tip is applied to the joint more times and for longer periods than when soldering.

Remember that components may be damaged by excessive heat, so space the desoldering operations out to allow the components to cool down in between.

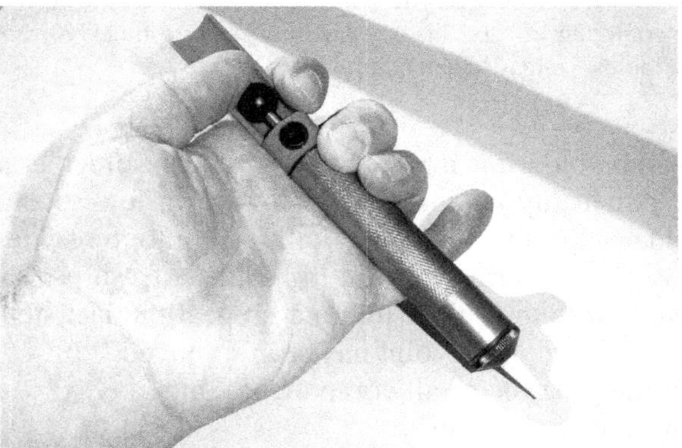

PHOTO 12.40 Priming the desolder pump by pressing down the charger.

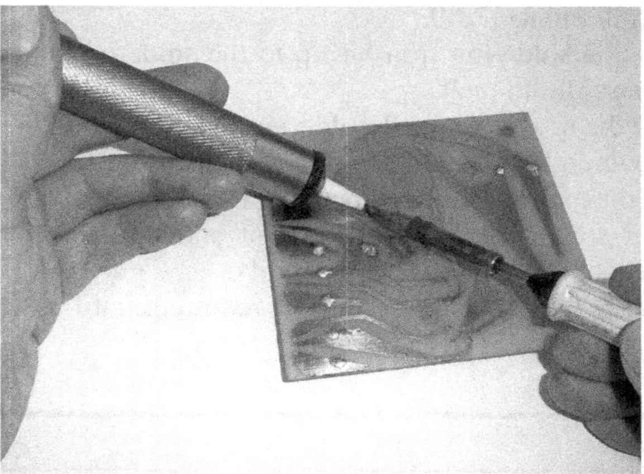

PHOTO 12.41 With the soldering iron bit tip heating the solder, apply the desoldering pump nozzle.

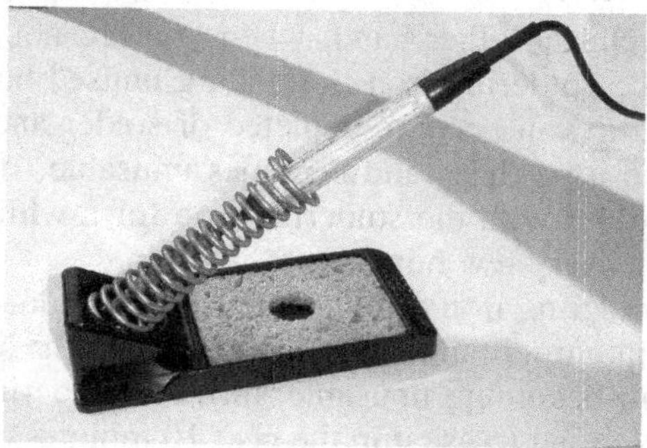

PHOTO 12.42 Use a soldering iron stand when not actually using your soldering iron. Keep the safety risks to a minimum.

Hint

You can now obtain soldering irons that come complete with an integrated desoldering pump. These can be very useful if you spend much of your time desoldering parts from printed circuit boards, say, if you worked in a service environment.

CARE OF YOUR SOLDERING IRON

Your soldering iron is a tool – a vital one when it comes to building electronic circuits, of course – and like any tool needs looking after, to keep it in good working order throughout its potentially long life. Things are made a little complicated in that the soldering iron is – I'll state the obvious here – very hot, so you also need to take certain safety precautions while using it. But, fortunately, on the other hand it's a fairly simple tool, so the things you need to do to look after it and the precautions you need to take while you're using it aren't excessive.

Looking After Your Soldering Iron

- Use a stand. When at rest, a soldering iron can be easily knocked or moved – it may fall off the work surface or simply touch against something else. Use a stand (as shown in Photo 12.42) when you are not actually soldering with it.
- Tin the soldering iron bit tip regularly – even at rest. When you solder with a soldering iron it is regularly in contact with fresh flux-cored solder, so the bit tip is maintained in good

condition. On the other hand, when you are not actually soldering and a soldering iron is standing unused but still turned on, the bit tip can become depleted of solder and flux due to the heat, so it oxidizes and becomes unusable. If you are not going to solder with the soldering iron for a while, remember still to tin it every few minutes.

- Turn the soldering iron off when unused. To reduce safety risks to an absolute minimum, and to help prevent the soldering iron bit tip from becoming unusable, turn the soldering iron off if you are not going to use it in the next 10 minutes or so.
- Clean the soldering iron bit tip when cold by lightly rubbing with a nylon pad. This may help a soldering iron bit tip that will not wet – but do not use wire wool or emery paper to clean the soldering iron bit tip, as this will remove protective plating and shorten the bit life.
- Never put a soldering iron – hot or cold – into liquid.
- Check the soldering iron's cable regularly for burns. A soldering iron when on is, of course, very hot, and accidental burning on the mains cable is a possibility.

PRINTED CIRCUIT BOARD LINKS

If a circuit is complex, it's probable that the printed board copper foil track layout will be complicated too. Under these circumstances, it's often the case that it's simply not possible to design a copper foil layout in a single layer. One trick to get around this problem is to use the component side of the board as well as the copper track side to complete the circuit connections. This is done with the use of simple links of wire, which connect from one point on the copper foil track, go over the track on the component side, then connect to another point on the copper foil track. Photo 12.43 shows an example of a printed circuit board that uses links.

Making a link:

- Cut a short length of single-strand wire. This should be a little longer than the distance the link needs to span between holes on the printed circuit board.
- Bend one end of the link around the nose of a pair of long-nosed pliers – see Photo 12.44. The bend should be at 90°, and there should be sufficient wire to go through the board and be soldered to a copper track.
- Measure the link length required.

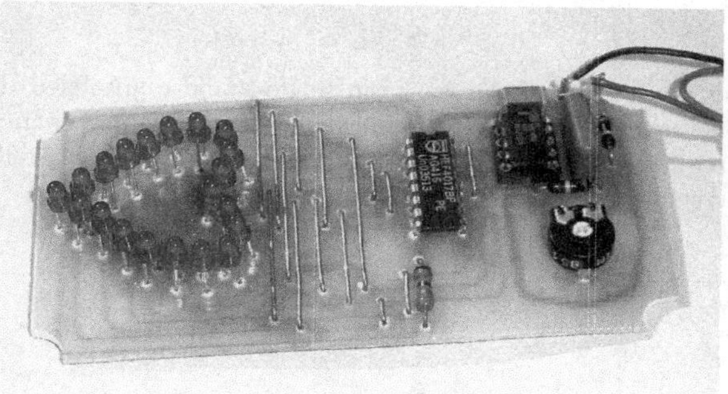

PHOTO 12.43 A printed circuit board that uses links to aid in copper foil track circuit layout. Note that many copper foil connections can take place under links, thereby easing a layout considerably.

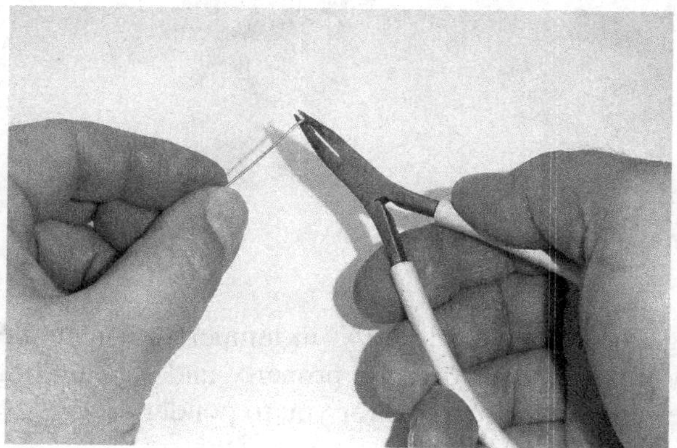

PHOTO 12.44 Bend one end of the link, using long-nosed pliers.

- Bend the other end of the link around the long-nosed pliers – see Photo 12.45. The length of link between bends should be the same as that you've just measured.
- Insert and solder the link.
- Trim excess link wire off the soldered joints.

SAFETY PRECAUTIONS

Soldering is a commonly used process, in industry, on the workbench, in the garage, in the home, which is generally hazard free. However, without correct and careful use of the tools involved, soldering can be very dangerous. When in use, the tip of a

Hint

We've shown a link being made from single-strand uninsulated wire, but such wires are – of course – er, uninsulated. If there is any chance that a short circuit may occur with other components or other wires, then use insulated single-strand wire to make your links.

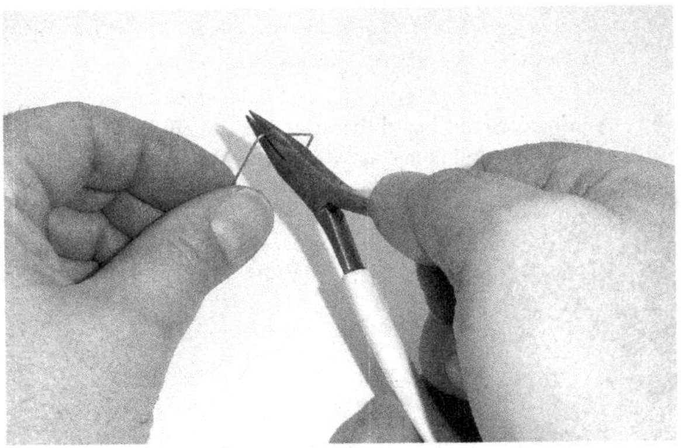

PHOTO 12.45 Bend the other end of the link.

soldering iron will exceed 400°C in temperature – ample heat to seriously injure you or damage property and surroundings. Here are some soldering safety tips for you to ponder:

- Always solder in a well-ventilated area – the fumes given off by molten solder can be irritating to respiratory membranes. Avoid breathing in fumes by tilting your head sideways, rather than holding it directly over your work. Wear a protective mask if necessary.
- Use a fire-resistant surface for your work. A small section of plasterboard is ideal, or you could invest in a special soldering mat built for the purpose.
- Always place a hot iron on a soldering iron stand in between use – never hang it over the edge of your bench or table.
- Replace your iron immediately if the electrical cord becomes frayed, worn, or otherwise compromised.
- Make sure the soldering iron is earthed as required.
- Never touch the element or bit of the soldering iron when it is in use, or for a while afterwards when it has been on.

- Wear heat-resistant gloves or use needle-nose pliers to grip components you're soldering.
- Components are very hot immediately after soldering, so let them cool before handling.
- Beware of splashes of molten solder from the soldering iron.
- Don't leave your soldering iron plugged in after you're finished using it. Never leave a hot soldering iron unattended.
- Beware of wire offcuts when snipping wires to length before or after soldering. They can fly with significant force, so protect your eyes.
- Always wash your hands after you finish using your soldering iron. Even if you use lead-free solder (lead can be toxic if ingested or absorbed into the skin) there are chemicals in the flux present in cored solder.
- Take great care to avoid touching the mains flex of the soldering iron with the element or bit of the iron. For extra protection, use a soldering iron that has a heatproof flex.

Having said all that, it is very seldom that soldering iron operators receive any burns or other injuries from the use of hot soldering irons. The technique is perfectly safe provided that common-sense precautions are taken.

FIRST AID

First aid when soldering usually amounts to burns treatments, and fortunately most burns from soldering are likely to be relatively minor, therefore treatment is simple:

- As quickly as possible, cool the affected area under gently running cold water. Keep the area in the cold water for at least 5 minutes (longer is recommended).
- Remove any objects or jewelry in case of swelling.
- Following cooling of the area in running cold water, do not apply creams or ointments, as the burn will heal better without them. A dry dressing may be applied to protect the area from dirt.
- Seek medical attention if required.

AND NOW THE TIME HAS COME …

So with this (fairly detailed, I must say) look at soldering, and the tools and processes it encompasses, thus ends our foray into the enjoyable world of electronics. You may feel as though you've learned a lot in the pages of this book and I trust you've had a good time in the process, but hopefully this is just the beginning of enjoyable times ahead.

Electronics is actually a much larger area than you've seen here, and there's much more to discover and much more to do. Indeed, you could spend literally years reading about electronics, and designing or building projects, and you still mightn't know it all but at least you are on the right track now – you're starting electronics … though just in case you think you already know it all, check out the quiz that follows!

QUIZ

Answers at the end of the book.

1. Soldering iron bits for electronics use reach a working temperature of:
   a. 400°C
   b. 282°C
   c. 100°C
   d. 1000°C.
2. Solder must contain at least a small amount of lead: true or false?
3. If you burn your finger on a soldering iron, you should immediately:
   a. Run the soldering iron bit under the hot tap
   b. Tap the soldering iron to teach it a lesson
   c. Turn the soldering iron off at the main supply
   d. Wrap the wound in a gauze bandage
   e. None of these
   f. a and d.
4. If you are called on to service old equipment and desolder components from printed circuit boards, you must check:
   a. If your soldering iron is RoHS compliant
   b. If your driving license is up to date
   c. If you have a gas mask in your toolkit
   d. If the equipment is RoHS compliant
   e. b and f
   f. None of these.
5. If you mistakenly solder a joint on a printed circuit board, desoldering tools such as desoldering braid or a desoldering

gun can always remove the solder, leaving the original copper surface as if untouched: true or false?

6. The junction between solder and copper in a soldered joint is called:
   a. Wet
   b. Lead-based
   c. An intermetallic bond
   d. RoHS compliant
   e. All of these
   f. c and d.
7. Before using cored solder in electronics, you should check that:
   a. You have pumped sufficient flux into the solder cores to finish the soldering job
   b. It is the correct sort (i.e. lead-based or lead-free) for the job
   c. You have sufficient gas in your soldering iron
   d. It is RoHS compliant
   e. It was made before 1 July 2006
   f. All of these
   g. None of these.
8. Place in order the following processes required to create a successful soldered joint:
   a. Trim component leads as required to just above the soldered joint
   b. Turn on the soldering iron to heat it to working temperature
   c. Apply solder to the joint
   d. Tin the soldering iron bit tip
   e. Preheat the joint
   f. Replace the soldering iron in its stand and allow the joint to cool.

Glossary

180° out of phase: When two identical sine waves are upside down with respect to each other.

Active: A device that controls current is said to be active.

Ampere: Standard unit to measure electrical current. Abbreviated to amp, symbol A.

Amplitude: The amplitude of an a.c. signal is the difference in voltage between its middle or common point and the peak voltage.

Analog: One of the two operating modes of a transistor, in which a variable tiny base current is used to control a large collector current.

Astable multi-vibrator: An oscillator whose output is a square wave.

Avalanche breakdown: The electronic breakdown of a diode when reverse biased to its breakdown voltage.

Avalanche point: The point on a diode's characteristic curve when the curve rapidly changes from a low reverse current to a high one.

Avalanche voltage: Breakdown voltage.

Axial: Component body shape in which connecting leads come from each of the two ends of a tubular body.

Base current: Shortened form for base-to-emitter current of a transistor.

Base, emitter, collector: The three terminals of a standard transistor.

Bias: To apply a fixed d.c. base current to a transistor, which forces the transistor to operate partially, even when no input signal is applied.

Bias current: The d.c. current applied to a transistor to force the transistor into partial operation at all times.

Bias resistor: In a common emitter circuit, the resistor that is connected to the base of the transistor, through which the bias current flows.

Block: To prevent d.c. signals passing through while allowing a.c. signal to pass – normally done with a capacitor.

Bode plots: Method of showing a circuit's frequency response, where straight lines are used to approximate the actual response.

Breadboard block: Tool that allows you to build circuits temporarily and test them. When you have completed your tests and are sure the circuit is working, you can remove the components and reuse them.

Breakdown: Effect in a diode, when reverse biased, where the reverse voltage remains more or less constant with different reverse currents.

Breakdown voltage: The reverse voltage that causes a diode to electronically break down.

Bridge rectifier: A collection of four diodes, either discrete or within an IC, which is used to provide full-wave rectification of an a.c. voltage input.

Buffer: (1) A device with high resistance input and low resistance output, which does not therefore load a preceding circuit, and is not loaded by a following circuit. (2) Another term for voltage follower.

Can: Nickname for the metal body of a transistor.

Capacitor: An electronic component consisting of two plates separated by an insulating layer, capable of storing electric charge.

Characteristic equation: The mathematical equation that defines the characteristic curve of an electronic component.

Circuit diagram: Method of illustrating a circuit using symbols, so that all electrical connections are shown but not physical ones.

Closed-loop gain: The gain of an operational amplifier with feedback.

Collector current: Shortened form for collector-to-emitter current of a transistor.

Conventional current: Current that is assumed to flow from a positive potential to a more negative potential. Conventional current is, in fact, made up of a flow of electrons (which are negative) from a negative to more positive potential.

Corner frequency: The frequency at which a signal size changes from one slope to another, when viewed as a graph of size against frequency. In the simple filter circuits in this book, the corner frequency, f, is given by the expression:

$$f = \frac{1}{RC}$$

Coulomb: Standard unit to measure quantity of electricity.

Current: Flow rate of electricity. Current is measured in amps.

Current gain: h_{fe}, shortened forms for forward current transfer ratio, common emitter, which is the ratio:

$$\frac{\text{Collector current}}{\text{Base current}}$$

of a transistor.

Decibels: Logarithmic units of gain ratio.

Dielectric: Correct term for the layer of insulating material between the two plates of a capacitor.

Digital: One of the two operating modes of a transistor, in which presence or lack of base current to the transistor is used to turn on or off the transistor's collector current.

Diode: A semiconductor electronic component with two electrodes, an anode and a cathode, which in essence allows current flow in only one direction (apart from when breakdown has occurred).

Diode characteristic curve: A graph of voltage across the current through a diode.

Discrete: Term implying a circuit built up from individual components.

Electrolytic capacitor: A capacitor whose function is due to an electrolytic process. It is therefore polarized and must be inserted into a circuit the correct way round.

Exponential curve: Capacitor charging and discharging curves are examples of exponential curves. An exponential curve rises/falls to about 0.63/0.37 of the total value in one time constant, and is within 1% of the final value after five time constants.

Farad (F): The unit of capacitance.

Feedback: When all or part of an op-amp's output is fed back to its input, to control the gain of the circuit.

Filter: A circuit that allows the signal of certain frequencies to pass through unaltered, while preventing passage of other signal frequencies.

Forward biased: A diode is forward biased when its anode is at a more positive potential than its cathode.

Frequency: The number of cycles of a periodic signal in a given time. Usually frequency is measured in cycles per second, or hertz (shortened to Hz).

Frequency response curves: Graphs of a circuit's gain against the frequency of the applied a.c. signal.

Full-wave rectification: Rectification of an a.c. voltage to d.c., where both half-waves of the a.c. wave are rectified.

Half-wave rectification: Rectification of an a.c. voltage to d.c., where only one half of the a.c. wave is rectified.

Hertz: The usual term for cycles per second.

High-pass filter: A circuit that allows signals of frequencies higher than the corner frequency to pass through unaltered, while preventing signals of frequencies lower than this from passing.

Hybrid transfer characteristic: A graph of collector current against base-to-emitter voltage (for a transistor in common emitter mode).

Input characteristic: A graph of base current against base-to-emitter voltage (for a transistor in common emitter mode).

Integrated circuit: A semiconductor device that contains a chip, comprising many transistors in a complex circuit.

Inverting amplifier: An op-amp with feedback, connected so that its input voltage is inverted and amplified.

Law of parallel resistors:

$$\frac{1}{R_{OV}} = \frac{1}{\text{R1}} + \frac{1}{\text{R2}} + \frac{1}{\text{R3}} + \ldots$$

Law of series resistors:

$$R_{OV} = \text{R1} + \text{R2} + \text{R3} + \ldots$$

Load: A circuit that has a relatively low resistance input is said to load a preceding circuit, by drawing too much current.

Load line: A line drawn on the same graph as a semiconductor's characteristic curve, representing the load's characteristic curve.

Low-pass filter: A circuit that allows the passage of signals with frequencies lower than the corner frequency, but prevents the passage of signals with frequencies higher than this.

Multimeter: A test meter that enables measurement of many things, e.g. voltage, current, resistance.

Non-inverting amplifier: An op-amp with feedback, connected so that its input voltage is amplified.

Offset null terminals: Terminals on an op-amp that may be used to reduce or eliminate the offset voltage present at the op-amp's output.

Offset voltage: The voltage present at an op-amp's output preventing it from equaling 0V when the input is 0V.

Ohmic: Term used to refer to a component whose characteristic curve follows Ohm's law.

Ohm's law: Law that defines the relationship between voltage, current, and resistance.

Op-amp: Abbreviation for operational amplifier.

Open-loop gain: An op-amp's gain without feedback.

Operating point: The point of intersection of a semiconductor's characteristic curve and the load line, representing the conditions within the circuit when operating.

Operational amplifier: A general-purpose amplifier (usually in integrated circuit form) that can be adapted and used in many circuits.

Oscillator: A circuit that produces an output signal of a repetitive form.

Output characteristic: A graph of collector current against collector-to-emitter voltage (for a transistor in common emitter mode).

Parallel: Joined at both ends.

Passive: A device through which current flows or doesn't flow, but which cannot control the size of the current, is said to be passive.

Peak voltage: The voltage measured at the peak of an a.c. signal.

Peak-to-peak voltage: The difference in voltage between the opposite peaks of an a.c. signal. Commonly shortened to p–p voltage or pk–pk voltage.

Periodic: Any a.c. signal that repeats itself regularly over a period of time is said to be periodic.

Permittivity (e): A ratio of capacitance against material thickness, of a capacitor dielectric.

Phase shifted: Two similar a.c. signals that are out of phase, i.e. start their respective cycles at different times, are said to be phase shifted.

Pointer: The indicator on a multimeter.

Potentiometer: A variable voltage divider used in electronic circuits. Many types exist.

Preset: A potentiometer that is set on manufacture and not normally readjusted.

Quiescent current: The standing current through a transistor due to the applied bias current.

Radial: A component body shape in which both connecting leads come out of one end of a tubular body.

Range: Measurement that a multimeter is set to read.

Reactance: A physical property of a capacitor that may be likened to a.c. resistance.

Relative permittivity: The ratio of how many times greater a material's permittivity is than that of air.

Relaxation oscillator: An oscillator relying on the principle of a charging and discharging capacitor.

Resistance: Property of a substance to resist the flow of current. Measured in ohms.

Resistance converter: Another term for voltage follower.

Resistor: Electronic component used to control electricity. A large number of different types and values are available.

Reverse biased: A diode is reverse biased when its cathode is at a more positive potential than its anode.

Ripple voltage: The small a.c. voltage superimposed on the large d.c. voltage supplied by a power supply.

Saturation reverse current: The small reverse current that occurs when a diode is reverse biased but not broken down.

Scale: The numbers marked on a multimeter that the pointer points to.

Self-bias: A form of biasing a common emitter transistor that regulates any variance in the transistor's current gain.

Series: Joined end to end, in line.

Sine wave: A constantly varying a.c. signal. Sine waves are generally used to test electronic circuits as they comprise a periodic signal of one signal frequency.

Smoothing: The process of averaging out a rectified d.c. wave, so that the resultant waveform is more nearly steady.

Square wave: A signal that oscillates between two fixed voltages.

Stabilize: To produce a fixed d.c. voltage from a smoothed one.

Three-rail power supply: A power supply that has a positive supply rail, a negative supply rail, and a 0V supply rail.

Time constant (*t*): The product of the capacitance and the resistance in a capacitor/resistor charging or discharging circuit.

Track: The fixed resistive part of a potentiometer.

Transfer characteristic: A graph of collector current against base current (for a transistor in common emitter mode).

Transistor: A three-layer semiconductor device, which may consist of a thin P-type layer sandwiched between two layers of N type (as in the NPN transistor), or a thin layer of N-type material between two layers of P type (as in the PNP transistor).

Transition voltage: The knee or sharp corner of a diode characteristic when forward biased.

Volt: Standard unit to measure electrical potential difference, symbol V.

Voltage: Potential difference – the flow pressure of electricity.

Voltage divider, potential divider: A number of electronic components in a network (usually two series resistors) that allow a reduction in voltage according to the voltage divider rule.

Voltage divider rule:

$$V_{out} = \frac{R2}{R1 + R2} \times V_{in}$$

Voltage follower: An op-amp circuit in which the op-amp is connected as a unity gain, non-inverting amplifier.

Voltage regulator: An integrated circuit that contains a stabilizing circuit.

Wavelength: The distance between two identical and consecutive points of a periodic waveform.

Wiper: That part of a potentiometer which is adjustable along the resistive track.

Zener diode: A special type of diode that exploits the breakdown effect of a diode when reverse biased.

Zero adjust knob: The control on an ohmmeter that allows the user to zero the multimeter.

Zeroing a multimeter: The adjustment made to an ohmmeter to take into account differences in voltage of the internal cell.

Component Suppliers (UK)

Over the next few pages is a list of some of the mail-order electronic component suppliers in the UK. The list is just a sample of suppliers, and the presence of any supplier on the list is by no means an endorsement.

Of course, one of the best places you can turn to these days to locate electronics components is the Internet – which is why I've included websites of the various suppliers. Use a search engine to locate more, and also you can look in your telephone book or Yellow Pages directory for local suppliers.

2001 Electronic Components Ltd
Eastman Way
Stevenage Business Park
Pin Green
Stevenage
SG1 4SU
Telephone: 01438 742001
URL: http://www.2k1.co.uk

Bardwells
288 Abbeydale Road
Sheffield
S7 1FL
URL: http://www.bardwells.co.uk

Bitsbox
41 Warwick Way
Olton
Solihull
B92 7HS
URL: http://www.bitsbox.co.uk

Bowood Electronics Limited
7 Bakewell Road
Baslow
Bakewell
Derbyshire
DE45 1RE
Telephone: 01246 200222
URL: http://www.bowood-electronics.co.uk

Cricklewood Electronics
40–42 Cricklewood Broadway
London
NW2 3ET
Telephone: 020 8452 0161
URL: http://www.cricklewoodelectronics.co.uk

Dannell Electronics Limited
Unit 2 Funston's Commercial Centre
Arkesden Road
Clavering
Saffron Walden
Essex
CB11 4QU
Telephone: 0845 603 1509
URL: http://www.dannell.co.uk

ESR Electronic Components
Station Road
Cullercoats
Tyne & Wear
NE30 4PQ
Telephone: 0191 2514363
URL: http://www.users.zetnet.co.uk/esr

Farnell InOne
Canal Road
Leeds
LS12 2TU
Telephone: 0844 711 1111
URL: http://uk.farnell.com

Fast Components Limited
Winchester House
Winchester Road
Walton on Thames
Surrey
KT12 2RH
Telephone: 0870 750 4468
URL: http://www.fastcomponents.co.uk

GreenWeld Limited
14 West Horndon Business Park
West Horndon
Brentwood
Essex
CM13 3XD
Telephone: 01277 811042
URL: http://www.greenweld.co.uk

Henrys Electronics Ltd
404 Edgware Road
Paddington
London
W2 1ED
Telephone: 020 7258 1831
URL: http://www.henrys.co.uk

JPR Electronics Ltd
4 Circle Business Centre
Blackburn Road
Dunstable
LU5 5DD
Telephone: 01582 470000
URL: http://www.jprelec.co.uk

Magenta Electronics Ltd
135 Hunter Street
Burton-on-Trent
DE14 2ST
Telephone: 01283 565435
URL: http://www.magenta2000.co.uk

Mainline Surplus Sales
Unit 1A Cutters Close Industrial Estate
Cutters Close
Narborough
Leicester
LE19 2FZ
Telephone: 0870 241 0810
URL: http://www.mainlinegroup.co.uk

Maplin Electronics Ltd
National Distribution Centre
Valley Road
Wombwell
Barnsley
South Yorkshire
S73 0BS
Telephone: 0870 429 6000
URL: http://www.maplin.co.uk

Mega Electronics
Mega House
Grip Industrial Estate
Linton
Cambridge
CB1 6NR
Telephone: 01223 893900
URL: http://www.megauk.com

Nikko Electronics
Dalbani House
358 Kingston Road
Ewell
Epsom
Surrey
KT19 0DT
Telephone: 020 8393 7774
URL: http://www.dalbani.co.uk

Quasar Electronics Ltd
PO Box 6935
Bishops Stortford
CM23 4WP
Telephone: 0870 246 1826
URL: http://www.quasarelectronics.com

Rapid Electronics Ltd
Severalls Lane
Colchester
Essex
CO4 5JS
Telephone: 01206 751166
URL: http://www.rapidonline.com

RS Components Ltd
Birchington Road
Corby
Northants
NN17 9R
Telephone: 01536 444222
URL: http://uk.rs-online.com

RSH Electronics
Unit 2
27A Ilkeston Road
Heanor
Derbyshire
DE75 7DT
URL: http://www.rshelectronics.co.uk

SIR-KIT Electronics
57 Severn Road
Clacton-on-Sea
Essex
CO15 3RB
URL: http://sir-kit.webs.com

Quiz Answers

Chapter 1:	1. d; 2. a; 3. e; 4. e; 5. c; 6. c
Chapter 2:	1. c; 2. a; 3. d; 4. True; 5. e
Chapter 3:	1. d; 2. e; 3. b; 4. True; 5. True; 6. a
Chapter 4:	1. a; 2. f; 3. c; 4. c; 5. c
Chapter 5:	1. a; 2. e; 3. c; 4. c
Chapter 6:	No quiz
Chapter 7:	1. d; 2. f; 3. c; 4. c; 5. True; 6. b
Chapter 8:	1. g; 2. a; 3. e; 4. b; 5. d
Chapter 9:	1. a; 2. f; 3. d; 4. e; 5. True; 6. False
Chapter 10:	1. b; 2. True; 3. e; 4. d; 5. e; 6. a; 7. c; 8. True
Chapter 11:	1. g; 2. True; 3. b; 4. d; 5. False; 6. a; 7. d; 8. b
Chapter 12:	1. a; 2. False; 3. e; 4. d; 5. False; 6. c; 7. b; 8. b, d, e, c, f, a

Index

C

H

I

J

K

L

T

Z

Printed in the United States
By Bookmasters